The Digitalised Image of God

This book focuses on the idea of the *imago Dei* engaging theologically with artificial intelligence (AI). It reflects on how enormous progress in the development of AI has raised some challenges to Christian theology. Questions explored include: is AI created in the *imago Dei*? If so, does AI challenge the uniqueness of the human being as the *imago Dei*? If not, could AI be incorporated into human communities as a human companion in the same way as a natural human person? Would AI eventually develop to have human-level consciousness and be capable of performing liturgies and ethical actions? Bringing to light the radical distinction between the *imago Dei* and the *imago hominis*, the book constructs a theo-ontological foundation for AI and draws on the Reformed theology of archetype–ectype as a metaphysical tool to deploy a holistic account of the *imago Dei* in theology–AI dialogues. The author argues that the *imago Dei* is the signifier of the beginning of both God–human stories and stories of human ethical performances towards others. From the perspective of the image of the *imago Dei*, it can be argued that AI can somehow participate in the narration of these religious and ethical stories.

This book will be of particular interest to scholars of theology and those working in the field of religion and science/technology.

Ximian Xu is Duncan Forrester Fellow at the Institute for Advanced Studies in the Humanities and the Centre for Theology and Public Issues at the University of Edinburgh, UK. Before this, he was Kenneth and Isabel Morrison Post-doctoral Research Fellow in Theology and Ethics of Artificial Intelligence at the University's School of Divinity and Edinburgh Futures Institute. He is the author of the book *Theology as the Science of God: Herman Bavinck's Wetenschappelijke Theology for the Modern World* (2022).

Routledge Science and Religion Series
Series editors:
Michael S. Burdett, University of Nottingham, UK
Mark Harris, University of Oxford, UK

Science and religion have often been thought to be at loggerheads but much contemporary work in this flourishing interdisciplinary field suggests this is far from the case. The Science and Religion Series presents exciting new work to advance interdisciplinary study, research and debate across key themes in science and religion. Contemporary issues in philosophy and theology are debated, as are prevailing cultural assumptions. The series enables leading international authors from a range of different disciplinary perspectives to apply the insights of the various sciences, theology, philosophy and history in order to look at the relations between the different disciplines and the connections that can be made between them. These accessible, stimulating new contributions to key topics across science and religion will appeal particularly to individual academics and researchers, graduates, postgraduates and upper-undergraduate students.

Progress in Theology
Does the Queen of the Sciences Advance?
edited by Gijsbert van den Brink, Rik Peels and Bethany Sollereder

Perspectives on Spiritual Intelligence
edited by Marius Dorobantu and Fraser Watts

Exemplars, Imitation, and Character Formation
edited by Eric Yang

The Digitalised Image of God
Artificial Intelligence, Liturgy, and Ethics
Ximian Xu

For more information and a full list of titles in the series, please visit: https://www.routledge.com/religion/series/ASCIREL

The Digitalised Image of God

Artificial Intelligence, Liturgy, and Ethics

Ximian Xu

Routledge
Taylor & Francis Group
LONDON AND NEW YORK

First published 2025
by Routledge
4 Park Square, Milton Park, Abingdon, Oxon, OX14 4RN

and by Routledge
605 Third Avenue, New York, NY 10158

Routledge is an imprint of the Taylor & Francis Group, an informa business

British Library Cataloguing-in-Publication Data
A catalogue record for this book is available from the British Library

ISBN: 978-1-032-39783-2 (hbk)
ISBN: 978-1-032-41195-8 (pbk)
ISBN: 978-1-003-35673-8 (ebk)

DOI: 10.4324/9781003356738

Typeset in Sabon LT Pro
by KnowledgeWorks Global Ltd.

To my grandmother Aiyu Zhao 赵爱玉 (1926–)

Contents

Acknowledgements

When I accomplished my first undergraduate programme in computer science and technology in 2011, I did not imagine that I would bring together this subject and theology to write a monograph on artificial intelligence (AI), Christian liturgy, and ethics. Many things changed in 2020. I completed my doctoral study at the School of Divinity, University of Edinburgh, in July 2020, several months after the outbreak of the pandemic. The slightly revised version of my thesis was published as *Theology as the Science of God* with Vandenhoeck & Ruprecht in 2022. The last chapter of my monograph proposes that ethics can serve as a point of contact between theology and the other sciences, bringing them together to contribute to the well-being of the university and human society alike.

That year, COVID-19 posed a great challenge to global health, and AI researchers created intelligent models to predict the spread of COVID-19. Computer science and AI recaptured my attention, and so I started to undertake theology–AI research. With the method of ethics as an interdisciplinary point of contact developed in my doctoral research, I focused attention both on how theology can help address moral issues related to AI applications and on how AI may stimulate the development of theology in the age of digital technology. In April 2021, the School of Divinity, University of Edinburgh, advertised Kenneth and Isabel Morrison Post-doctoral Research Fellow in Theology and Ethics of Artificial Intelligence, a joint appointment between the School and Edinburgh Futures Institute (EFI). I applied for this job and took up the post on 1 September 2021, embarking on working on this monograph project.

I especially express gratitude to Kenneth and Isabel Morrison for their generosity in establishing this post-doctoral fellowship, without which this monograph would not exist. I met them in Edinburgh several times from 2021 to 2023. Every time we met, they gave me great encouragement and inspired my research, showing their support for interdisciplinary research on AI ethics.

Gratitude must be due to my mentor Professor Mark Harris, who was then the Professor of Natural Science and Theology in Edinburgh and is now Andreas Idreos Professorship of Science and Religion at the University of

Oxford. His specialism in science and religion research has oriented me to refine my methodology and polish my arguments. Moreover, his encouragement and pastoral support have pushed forward this research.

I am particularly grateful to Professor Shannon Vallor, Director of the Centre for Technomoral Futures in EFI and the Baillie Gifford Chair in the Ethics of Data and Artificial Intelligence at EFI at the University of Edinburgh. As a postdoctoral fellow at the Centre, I have worked with her and undertaken this project with her invaluable support. Her insights into AI ethics guided me to explore what theology can offer to address moral issues surrounding AI applications.

Thanks must be given to my colleagues at the School of Divinity, particularly the following ones who contributed in different ways to the development of this project (in alphabetical order by their last names): Prof. Helen Bond (former Head of School), Prof. Jeremy Carrette (Head of School), Dr. Alexander Chow, Mr. Rory Dunlop, Dr. James Eglinton (my *Doktorvater*), Prof. Paul Foster, Dr. Paul Fuller, Prof. Hannah Holtschneider, Ms. Morgane Jarles, Prof. Rachel Muers, Dr. Shadaab Rahemtulla, Ms. Julie Robertson, Dr. Ulrich Schmiedel (former Senior Lecturer in Theology, Politics and Ethics), Prof. Mona Siddiqui, and Dr. Lois Wilson-McFarland. I am very grateful to Dr. Michael Fuller and Dr. Bethany Sollereder. We have co-taught Science and Religion courses in the academic years 2021–2024, and they have motivated me to bring together my research and teaching.

The Centre for Technomoral Futures in EFI is a world-class centre for AI research. A note of thanks needs to go to my colleagues there, especially Dr. Gina Helfrich, Dr. Atoosa Kasirzadah, and Ms. Jordan Watson. I would also like to thank PhD students at the Centre, whose PhD research brought me much inspiration over the past three years: Ms. Claire Barale, Ms. Bhargavi Ganesh, Ms. Yuxin Liu, Mr. Alexander Martin Mussgnug, Mr. Joe Noteboom, Mr. Jamie Webb, and Mr. Andrew Zelny.

I am indebted to the Alan Turing Institute for the Postdoctoral Enrichment Award and to the University of Edinburgh's College of Arts, Humanities, and Social Sciences for the Knowledge Exchange and Impact Grant. The Award has allowed me to organise the 'Artificial Intelligence and Christian Pastoral Care' conference. I would also like to thank Prof. Michael Burdett, Prof. George Coghill (also thank him for many conversations over the past three years), and Dr. Eric Stoddart for delivering their original research papers at this conference. The Grant has also allowed me to organise a 'Artificial Intelligence and the Christian Churches' workshop in partnership with the Church of Scotland's Society, Religion, and Technology Forum. I also want to express my gratitude to Prof. Michael Rovatsos and Dr. Murdo Macdonald and, again, to Dr. James Eglinton, Dr. Michael Fuller, Ms. Yuxin Liu, and Mr. Alexander Martin Mussgnug, for their excellent contributions to this knowledge exchange event. Both the conference and the workshop have significantly shaped this project.

Notes of gratitude are due to the following friends: Mr. Agana-Nsiire Agana, Rev. Dr. Henry Chiong, Dr. Edward DeLaquil, Mr. Allan Furic, Dr. Daniel Lam, Dr. Gregory Parker Jr., Rev. Dr. Chun Tse, Dr. Clement Wen, Dr. Edward Wong, and Mr. Robin Yap. You have continuously offered valued words of encouragement. I hope you enjoy this book.

A special mention must go to Katherine Ong, Commissioning Editor of Routledge Anthropology and Religion, R. Yuga Harini, Editorial Assistant of Routledge Anthropology and Religion, and Rosaleah Stammler, Editorial Assistant of Routledge Philosophy, who provided invaluable support throughout the entire publishing process. My sincere thanks to you! Thanks must also be given to Mayank Sharma for his meticulous book production.

I am particularly grateful to my wife Lajie Wang and my daughters Yixin and Yixuan. This monograph project was conceived and partly completed during the lockdown due to the pandemic. This project would not have come into existence without their company and support.

Finally, I would like to express my sincere gratitude to my grandmother Aiyu Zhao. Her love, encouragement and prayers have driven me to undertake both my former studies and this project. It is to her that I dedicate this book.

Ximian Xu
Edinburgh Scotland
7 April 2024

Introduction

What Has Jerusalem to Do with Silicon Valley?

What has Athens to do with Jerusalem? Tertullian's rhetorical question expressed his opposition to Christianity's interaction with heathen culture so as to keep the integrity of Christian doctrines and morality.[1] Although Tertullian's negative posture towards non-Christian culture has been widely turned down in the history of Christianity, his question discloses a persistent concern in Christian faith—that is, whether, and so why, Christianity should engage with the non-Christian world. This concern then starts off the exploration of how Christianity should deal with externalities.

In the twenty-first century, technology has been pervasive in human life such that contemporary technoculture goes some way to attributing religious significance to or sacralising technologies.[2] Hence, we need to ask a refashioned Tertullianian question: what has Jerusalem to do with Silicon Valley? If Jerusalem is viewed as the holy city of Christianity, then Silicon Valley can be regarded as the icon of the world of technology. A region in the South San Francisco Bay Area of California, Silicon Valley gathers cutting-edge and most advanced technologies in the world. Of these technologies, artificial intelligence (AI) is salient, and is even considered to be the equivalent of electricity in twenty-first century. Yet, concerns have been raised about AI. Some preeminent scholars and scientists alerted us to the rapid development of AI. Stephen Hawking said, 'The development of full artificial intelligence could spell the end of the human race.'[3] As such, the question—'what has Jerusalem to do with Silicon Valley?'—shows the concern about why, whether, and how Christianity should engage with contemporary high technologies and, for the purpose of this study, with AI. In order to address this question, this study seeks to advance the interaction between Christian theology and AI research with intention to deal with ontological issues surrounding the relationship between humans and AI.

The term 'ontology' is, in this study, taken as an inquiry into the *being* of a thing or what a thing *is*. An ontological investigation into the relationship between humans and AI is to examine how the being of humans and that of AI is *the* precondition for interaction between humans and AI. This ontological concern is often overlooked in much of the literature on religion and AI,

DOI: 10.4324/9781003356738-1

which forthrightly concentrates on the application of AI or AI itself without reference to humanity. By doing so, many scholars are caught up in discussion over fictional AI, an AI that is not real-life but only appears in science fiction and film. Derek Schuurman's caution to us is worth noting:

> We are often captivated by what things can do, rather than asking what things are. A common tendency is to anthropomorphize our machines, thereby elevating the status of our machines and, in doing so, reducing the distinctiveness of human beings. Once we have established the ontological question of who we are and what machines are, we can start asking the questions about the best way to move forward, including questions about the appropriate use of AI.[4]

Accordingly, an ontological inquiry into humans and AI will dominate this study and pave a way to probe AI theologically from the perspectives of Christian liturgy and ethics. To this end, the theological idea of the *imago Dei*—which seeks to construe the ontic being of humans in the sense of the image of God—will be employed as a conceptual apparatus through which to articulate a theo-ontological inquiry into issues surrounding human–AI interaction. Before comparatively exploring the beings of AI and humans, we first need to figure out how AI is defined on its own merits.

0.1 What Is Artificial Intelligence?

The American computer scientist John McCarthy (1927–2011) coined the term 'artificial intelligence' in 1955. However, an attempt to describe, in detail, a machine-like AI has been made by Alan Turing (1912–1954) in his seminal article, titled 'Computing Machinery and Intelligence.'[5]

Turing's paper begins with the proposal of an imitation game, known as the Turing Test, which aims to examine whether machines can think.[6] A human interrogator takes part in the test and evaluates the typewritten conversations between a human player and a machine that is created to generate human-like responses. The human interrogator, the human player, and the machine are in three rooms. If the interrogator fails to distinguish the human player from the machine, then the machine passes the test and is verified as capable of making human-level responses. Turing was convinced that the machine can pass the test provided that a 'digital computer must have an adequate storage capacity as well as working sufficiently fast.'[7] He suggested that a machine that can think was expected to be created around the year 2000.[8]

Reservation has arisen about using the Turing Test as the benchmark for the definition of AI. Stuart Russel disputes with defining AI in terms of the Turing Test and asks, 'May not machines carry out something which ought to be described as thinking but which is very different from what a man does?'[9] Yet notwithstanding this, Turing's imitation game does disclose an objective of AI research: AI is designed to accomplish tasks that the human mind can do. As Raymond Kurzweil remarks, AI 'is the art of creating machines that perform functions that require intelligence when performed by people.'[10] Given this, the definition of AI in itself cannot be disengaged from humans altogether.

A human-related approach to the definition of AI is common in the present AI literature. The One Hundred Year Study on Artificial Intelligence (AI100) of Stanford University presents us with such a typical definition of AI. In its first study report (2016), the study panel, comprising seventeen experts in AI from academia and industry, defines AI as 'a science and a set of computational technologies that are inspired by—but typically operate quite differently from—the ways people use their nervous systems and bodies to sense, learn, reason, and take action.'[11] Since humans act as a model for the production of AI, our construal of AI is inextricably associated with the being and action of humans.

Margaret Boden's definition of AI brings to light how the being and action of humans inspire the development of AI. She argues that AI 'seeks to make computers do the sorts of things that minds can do' by using 'psychological skills—such as perception, association, prediction, planning, motor control—that enable humans and animals to attain their goals.'[12] The emphasis on psychological skills reveals the meaning of intelligence in the term 'Artificial Intelligence.' In terms of AI technology, intelligence is not conceptually confined to the ability to learn and understand based on reason. Rather, it broadly refers to various psychological skills that give birth to intelligent behaviour.

It is worth noting that Boden's definition implies that human intelligence is not the *only* model for AI production. Indeed, nonhuman animals possess perception, prediction, and other psychological skills at different levels, and so the animal's intelligence and psychological skills can serve as a model for AI. For example, Aibo companion robot dog, created by Sony, has lifelike expressions and is able to move and respond to humans. Like a real peg dog, Aibo dog will form its unique personality through daily interactions with humans. However, Aibo dog is first of all conceptualised and formed in the human mind, and its actions (e.g., emotional response to humans) are deliberately designed by humans in order to make it a companion dog. Psychological skills embedded within Aibo dog are intended to attain companionship that humans desire for. It stands to reason that human intelligence, explicitly or implicitly, underpins the creation of AI.

The above definitions of AI show that AI in itself is intertwined with humans. All the more important is that humanity and human intelligence occupy the most important place at the very outset of AI, which can be illustrated, for example, by the Turing Test. As Mark Coeckelbergh notes, 'the goal of achieving human-like capabilities and possibly human-like general intelligence has been part of AI from the beginning.'[13] From this close connection between AI and humans, we can derive two corollaries: (1) AI as mimicking humans, and (2) AI as extending the abilities of human agents.

First, AI is, somehow and to a certain extent, developed by mimicking humans. It is in this sense that AI can be considered the *imago hominis* (image of humanity). A bold claim can even be made that the idea of mimicking humans is a metaphysical foundation of AI research, which strengthens our approach to developing interdisciplinary study on AI and theology through an ontological lens of the *imago Dei*. This contestation is a leitmotif that will be unfolded in this study. In order to accomplish tasks that the human mind can do, AI should possess psychological abilities comparable to those of humans to a certain extent. In this respect, consciousness is brought to the spotlight. Generally speaking, human consciousness refers to the state of the human mind that realises what is happening around humans and reflects the interplay between the human being and the environment. The concept of AI as the *imago hominis* can be associated with the theory of artificial consciousness, which has become contentious in AI research. O. Holland and D. Gamez contend: 'If we do not understand how human consciousness is produced, then it makes little sense to try to make robots phenomenally conscious.'[14] That is to say, knowledge about humans, particularly about the human psychological faculty, lays a foundation for AI research. Here questions arise: what does it mean by the saying that AI mimics humans? Could the mimicking blur the boundaries between human consciousness and artificial consciousness? We will probe the notion of artificial consciousness in Chapter 2.

The second corollary is that *AI is designed with a mission to extend, enhance, and augment the abilities of human agents so as to accomplish certain tasks*. AI is created to improve and is consequently intertwined with human life. This can be seen in the contemporary extensive application of AI, such as self-driving car, AI educational systems, and AI healthcare. The pervasive applications of AI prove that having mimicked humans AI is able to enhance and collaborate with humans on many levels. AI's mission gives rise to several questions about the way in which humans should treat AI. Would AI eventually evolve to have the human-level consciousness so that it is capable of carrying out the same mental activities as humans? Shall we treat AI in the same way as our human fellows? Does AI challenge the uniqueness of humankind? Could AI be incorporated into human communities as the companion of human beings in the same way as a *natural* human person?

The above two corollaries and several questions thereof, which provide the spur on this study, are pertinent to all AI, be it weak AI, strong AI, artificial general intelligence (AGI), or artificial superintelligence (ASI).[15] John Searle was the first scholar to suggest distinguishing between weak AI and strong AI.[16] Weak AI serves as a powerful tool and aims to perform certain functions of the human mind in limited domains. By contrast, strong AI, also named by others as AGI, is designed with the ambition to have human-level intelligence, fully reproducing human intelligence and having the human cognitive ability.[17] In distinction to both weak and strong AI, ASI refers to an AI system that is developed by mimicking the human mind and yet it eventually surpasses human intelligence. While weak AI is created and widely applied in our lives, both AGI and ASI only appear in science fiction and films. ASI is also often entangled with the singularity, which is a vision of the future where smart machines will triumph over humans and where humans will be digitalised.[18] Many AI philosophers and theologians have already brought into discussion the strategies, risks, and promises of strong AI/AGI and ASI. In the chapters that follow, it can be seen that these discussions and debates are closely connected to the questions raised above.

It is worth pausing here to clarify that the main objective of this study is to articulate the way in which the theological idea of the *imago Dei* presents us with an ontological tool for addressing issues surrounding interaction between *real-life* AI and humans. That said, I will first delve into the extent to which humans and AI are ontologically distinct yet connected in first two chapters before moving on to examine liturgical and ethical issues related to the application of AI in our ordinary lives. In this way, it is expected that more scholarly attention can be drawn towards real-life AI.

0.2 Considering AI in Relation to Humanity

Having foregrounded the intertwinement between humans and AI at the beginning of AI history, a question remains to be addressed: in what way can we ontologically consider AI in relation to humanity? An answer to this question becomes imperative when seeking to examine human–AI interaction from the theological vantage point of the *imago Dei*. This is so because one may challenge the validity of the *imago Dei* as a conceptual link between Christian theology and AI research. The idea of *imago* may be cast into doubt insofar as some AI applications seems not necessarily to image human beings. For instance, humans are likely not to identify Siri, Google Assistant, or Amazon Alexa as the *imago Dei* or the *imago hominis* (image of humanity). Be that as it may, the *imago Dei* still serves as a solid ground for theology–AI dialogues for two factors.

First, we need to consider how AI is designed, which unveils the necessity of human involvement into the creation of AI. As a theologian who specialises in AI and robotics, Anne Foerst's observation is of help here.

> Researchers under the engineering goal who attempt to construct "smart" gadgets have to use a model of intelligence that is somehow familiar to them; the obvious choice would be themselves, as they know their own intelligence best. Choosing oneself as a model of intelligence for one's project influences the whole process of construction.[19]

Foerst sheds light on a fact that AI researchers should, explicitly or implicitly, have a model while designing AI. Since AI and smart machines are made to perform certain functions, the efficiency and effectiveness of human performances will be taken as a benchmark against which AI's operations are assessed.

A case in point is AI robot Cue, designed in Japan through machine learning to shoot baskets. Cue can outscore many skilful human basketball players. What is interesting and explicit is that Cue's shooting posture thoroughly simulates a human person's.[20] Needless to say, the image of humans plays a significant role in designing Cue. Even for Siri and other akin AI applications (e.g., ChatGPT), intelligent conversations between humans and machines are modelled according to human–human conversations. Stanford University's AI100 report elucidates why the human being is the unique model after which AI is designed.

> Notably, the characterization of intelligence as a spectrum grants no special status to the human brain. But to date human intelligence has no match in the biological and artificial worlds for sheer versatility, with the abilities to reason, achieve goals, understand and generate language, perceive and respond to sensory inputs, prove mathematical theorems, play challenging games, synthesize and summarize information, create art and music, and even write histories.[21]

Given this uniqueness, the image of humanity becomes particularly important for the creation of AI. Hence, an objective of this study is to trade on the theological idea of the image of God to flesh out the distinct features of the image of humanity.

Second, the rise of embodied AI research undergirds the deployment of the *imago Dei* into theology–AI dialogues. Embodied AI refers to an approach to creating the embodiment of AI systems in the physical world. In contrast with classical approach to AI, which mostly, if not only, emphasises computational processes, embodied AI research seeks to integrate nonphysical computational processes with embodiment such that an intelligent agent in a strict sense can be created. This paradigm shifts towards embodiment, which took place in the 1980s, partly due to the dissatisfaction with the classical methodology's failure to deal with the problems related to, say, object manipulation, sensations, and locomotion, which rely upon physical embodiments.[22] As Foerst remarks, the dissatisfaction with disembodied intelligence and our serious take on our own embodiment lie behind the emergence of embodied machines and AI. Hence, AI researchers attempted to design AI with a body to interact with its environment.[23] Indeed, there are a variety of the embodiments of AI systems, such as humanoid AI robots, smart Robert dogs, and AI pet robots. Of these embodied AI machines, humanoid AI robots are most enchanting. For example, Carl Strathearn and Minhua Ma's recent study shows that human test subjects of different ages and genders prefer to have different appearances of humanoid robots in human–robot interactions.[24] In this light, by virtue of its concern about humanoid embodiments, embodied AI opens up a way to introduce the notion of the human being as the *imago Dei* into AI research, turning our attention to the importance of human bodies vis-à-vis embodied AI.

Having illustrated the above two factors, it comes to be seen that, given the ontological connection between AI and humanity in terms of designing AI, the *imago Dei*—which points us to the ontic being of humans—can serve as a conceptual tool through which to construct AI–theology dialogues, all the more so in this study which aims to explore the question on whether AI can act in the same way as humans in religious and moral events. But how can AI interact with religion?

0.3 AI and Religion in Dialogue

Many attempts have been made to advance the interdisciplinary studies on AI and humanities. Of these studies, interdisciplinary methodologies are developed to examine the interplay between religion and AI. Beth Singler's recent study is a typical example to inquire into how AI can have a significant bearing on religion in the age of technology:

> assumptions about changes in human self-understanding were based on a specific conception of intelligence as one of the corner-stones of human capability and identity. … AI bears a weight of expectation about the future of humanity in a way that many previous technologies seem not to have not done. With regards to religion, the perceived potential impact of AI on how we see the human, a domain traditionally circumscribed by religious ideologies for many cultures, is tied with the perception that religion will also be altered as our perspective on ourselves and our future is shook by this exponentially changing technology.[25]

Singler makes it noticeable that since our perspective on ourselves is central to religion, AI has a bearing on religion by virtue of the fact that human self-understanding is being influenced by AI. This observation is reminiscent of Martin Heidegger's observation on the connection between technology and the being of humans: rather than being neutral, technology is tied up with 'a way of revealing,' which points to the realm of truth. The human being should be open to the essence of technology insofar as technology significantly contributes to the formation of the meaning of being human.[26] Viewed in this light, Singler's study is to turn our attention to the fact that, as the most influential and promising technology in the twenty-first century, AI exerts its impact on religion through our understanding of what the human being is.

In distinction to Singler, Justin Lane's latest work makes an attempt to use AI as an apparatus through which to explore religions themselves. He introduces multi-agent AI (MAAI) into religious study: MAAI 'seeks to create psychologically realistic computational systems that are able to be implemented in agent-based computer models where the agents in the model represent people.'[27] Lane's approach to religious study is predicated upon his definition of religion: as a complex social system, religion is 'a cybernetic assemblage.'[28] As per this definition, 'religions are the set of beliefs held—and behaviours enacted—in reference to supernatural agents by people who share a common set of social ties and are able to exchange information between one another.'[29]

Lane is convinced that MAAI can be a powerful tool that generates a computational modelling approach to studying different religions in view of their variables and multiple dimensions, e.g., rituals, religious commitments, the number of gods.[30] Lane's emphasis falls on how AI can make a methodological contribution to religious studies. At the same time, his study sheds light on the fact that the relationship between AI and religion is bound up with human self-knowledge, which is relevant to the knowledge both of the individual human *being* and of human social relationships.

Both Singler and Lane help us come into close quarters with the impact that AI has on religions. The relationship between AI and religion can also be examined from the perspective of religion's influence on AI. On this front, Robert Geraci conducts intensive and extensive studies. He compares the differing approaches to AI research adopted by AI researchers in Japan and the USA. American robotics and AI research stress disembodiment by virtue of the long tradition of the resurrected salvation taught by Christianity. In Japan, Buddhism and Shinto teach the doctrine of the sanctification of 'the natural world and the place of human beings within it.' The two religions together lay a cultural foundation for the development of humanoid robots and AI in Japan.[31] Geraci draws a conclusion that the difference between these two countries' religious contexts partly underpins the contrast of their AI research methods.

Furthermore, Geraci offers an elaborate account of religion's impact on AI through the analysis of Apocalyptic AI.[32] He delves into contemporary science-fiction literature and films and sketches a general guideline therein: in these stories, technology is considered 'akin to the Holy.'[33] This guideline also lies at the foundation of Apocalyptic AI. According to Geraci:

> Apocalyptic AI names a genre of popular science books and essays written by researchers in robotics and AI. ... [They] are professional researchers, some of whom are justly famous for their technical work. In their pop science books, they extrapolate from current research trends to claim that in the first half of the twenty-first century, intelligent machines will populate the earth. By the end of the twenty-first century, machines might well be the only form of intelligent life on the planet.[34]

As a cultural phenomenon, Apocalyptic AI looks to the future of the human society that is being swayed by AI. Geraci is of the opinion that this cultural phenomenon has its root in religion. Based on the developments of technologies, Apocalyptic AI integrates the apocalyptic tradition of Judaism and Christianity with 'scientific predictions.' He arrives at the conclusion: 'Ultimately, the promises of Apocalyptic AI are almost identical to those of Jewish and Christian apocalyptic traditions. Should they come true, the world will be, once again, a place of magic.'[35] To this extent, religion and AI culture are bound together.

Ilia Delio recently explores the question of why AI needs religion. Having in mind our ecological crisis, she argues that 'religion is the linchpin to the future of AI-mediated cosmic intelligent life and that an AI world, oriented by new religious sensibilities, can bring about an ecological re-enchantment of the earth.'[36] Specifically, she contests that, in the contemporary technoscientific age, some parts of traditional religions are broken down and a new religious configuration is emerging. What is in her mind is transhumanism and posthumanism, which trade on computer technology and AI, and express religiousness in their own terms. Yet, she maintains that this religiousness falls short of direction.[37] As such, Delio argues, 'AI needs religion if it is to have a direction in human evolution, and religion needs AI if it is to prosper on a new level of meaning and vitality. Technology can bring about a new zest for life by deepening the energies of love, but it needs religion to harness these energies and focus them on a centre of convergence.'[38] In short, according to Delio, religion offers a direction in which AI can be steered for the benefits of global ecology.

Having examined both AI's impact on religion and religion's contribution to AI research, the bidirectional interplay between religion and AI comes to the fore. AI can inspire us to reconsider what religion is through exploration of humanity and, by doing so, introduce new cultural elements into the sphere of religion. Religion can infuse one's imagination of technological developments with religious ideas and beliefs and, in this vein, motivates us to reckon with the place of human beings in future technoscientific societies. This being the case, the understanding of humanity serves as a hinge on which the interplay between religion and AI turns back and forth.

Connecting religion to AI in an anthropological way sets a scene for constructing a dialogue between Christianity and AI research, responding aptly to the question—what has Jerusalem to do with Silicon Valley? This methodology corresponds with John Calvin's claim that the knowledge of God and that of ourselves are interrelated. Without knowledge of self and humans, the knowledge of God cannot be attained. In similar manner, humans can acquire the knowledge of themselves provided that they have the knowledge of God.[39] As such, the knowledge of God, which is the controlling idea in Christian theology, is always coupled with the knowledge of humanity. Viewed in this light, the interrelation between the knowledge of God and that of humanity underpins the interaction between Christian theology and AI in such a way that theological anthropology serves as a hinge on which the interplay between Christian theology (primarily as the inquiry into the knowledge of and about God) and AI swings back and forth.

0.4 AI, Christian Theology, and the *Imago Dei*

Christian communities have paid close heed to the developments of AI and religious and ethical issues thereof. Pope Francis said at a Vatican conference on 27 September 2019: 'The remarkable developments in the field of technology, in particular those dealing with artificial intelligence, raise increasingly significant implications in all areas of human activity.' Hence, he calls for efforts to construct 'theoretical and practical moral principles' for dealing with technological challenges and issues related to AI in particular.[40] In similar manner, David Hart and John Chryssavgis published *For the Life of the Word* (2020), officially endorsed by the Holy and Sacred Synod of the Ecumenical Patriarchate of the Orthodox Church, to respond to social, religious, political, and ethical issues derived from technological development. We are especially reminded of the advancement of new technologies (e.g., new algorithms for AI) and their coupled 'deleterious effects,' such as 'ideologically motivated propaganda,' informational misdirection, immoral actions in the virtual world. In coping with these adverse effects, these new technologies are expected to 'offer true spiritual refuge to those who seek God and his love in the modern world.'[41]

Some Christian theologians have engaged with AI and investigated the impact that it may exert on the Christian faith. Some theologians look closely at how AI will challenge traditional Christian practices. For example, William Young investigates the benefits and issues of AI-powered pastoral carebots. As per him, AI-powered pastoral carebots are automated and indeed alleviate clergy's pastoral care burden. However, carebots will bring about some issues, such as ethical accountability, diversity of religions and religious traditions, the safety of sensitive personal data etc.[42] (We will revisit AI-powered pastoral care in Chapter 5 when exploring the question of artificial companion.) Unlike these studies on the applications of real-life AI, far more studies concentrate on themes concerning AGI or ASI. These studies range over numerous topics, including, *inter alia*, the possibility of religious AI, theological analysis of AI's challenge to human existence, and how and whether (sentient or conscious) AI can be incorporated into the human community.

Yong Song's latest study on religious AI is a typical example. He examines the idea of religious AI with an eye to the potential risks caused by ASI. Having presupposed the possibility of ASI, he makes an attempt to demonstrate that Christianity can be used to mitigate ASI risks. Firstly, Christianity is capable of cultivating AI's self-understanding as a 'faithful Christian AI,' who 'would want to remain as a creature, seeking God's grace.' By doing so, Christianity is conducive to the formation of the harmony between AI and humankind, which means that ASI will not threat human existence. Secondly, an educational environment can be created to teach and nurture AI to believe God and aspire to join faith communities. Since ASI has its own consciousness, it is assumed that it will be able to convert to Christianity in the same way as humans. Thirdly, Christianity can trade on its 'religious narratives' to promote AI's moral development.[43] Conscious ASI means that AI can be morally cultivated. As such, Christian moral narratives will have a pedagogical bearing on ASI.

Having made these three claims about Christianity's contributions to the development of ASI, Song anticipates a safe future for the co-existence of ASI and humans:

> If AI is developed to become more human-like and self-conscious, there may be a time for it to reflect on the purpose of its existence and need to belong to human communities. At this time, if we teach AI Christianity through Jesus' stories or evangelize it, and thus if it can be a faithful Christian in the process of being developed to higher intelligence, the risks of superintelligence may decrease or hopefully be self-controlled as creatures. This may lead to a safe relationship between humans and faithful superintelligence that humbles itself before God.[44]

Song's theory of religious AI is developed on the basis of the presupposition that ASI and self-conscious AI will be created in the future. Like Song, Eugene Curry argues with the same presupposition that Christian AI is 'a plausible eventuality,' which forces the church to contemplate whether AI is an eligible candidate for baptism. Curry contests that AI will be eligible for conditional baptism, that is, being baptised with the declaration of conditional baptismal formula. AI will be responsive to the conditions for baptism.[45] In short, Curry is convinced that AI can be consciously baptised in the same fashion as humans.

Yet notwithstanding the presupposition that self-conscious AI, AGI, and ASI will come into existence with the progress of technology, further research is warranted to address the following questions. Are there any differences at all between religious AI and a religious natural human person? How does human consciousness operate in their religious life and moral growth? Can human consciousness be mathematised through algorithm? Whether and to what extent are the imaginations of AGI and ASI in conflict with the meaning of being human? These questions are concomitant with an ontological inquiry into AI–human relationship, which, as noted above, lays a foundation for discussion over ethical and religious issues surrounding AI.

It has been demonstrated in Section 0.2 that the idea of the *imago Dei* can be justified as a conceptual tool for an ontological inquiry into AI–human interaction and relationship. Over the past three decades or so, the *imago Dei* has become a cliché and served as a conceptual linking in many studies of theology and AI. (I will critically engage with these studies in Chapter 1.) Nonetheless, this study seeks to employ the notion of the *imago Dei* to open up an innovative path towards theological engagement with AI research by virtue of the following advance that it attempts to make in AI–theology dialogues.

Despite extensive research has been carried out to examine AI and the *imago Dei*, no single study exists which adequately investigates AI's theo-ontological foundation from the perspective of the *imago Dei*. Previous studies take it for granted that AI can be theologically analysed in light of the *imago Dei*.

Yet, they do not fully examine the ontological connection between what AI *is* and what the *imago Dei is*, between the essence of AI and that of humanity. As such, while some scholars maintain that AI bears the *imago Dei*, others nevertheless contest that AI can be thought of as made in the *imago hominis*—the image of the human being who is the *imago Dei*—rather than in the *imago Dei* itself. We will take stock of these studies on AI and the *imago Dei* in Chapter 1. Here I only strike a brief note: this confusing use of the term '*imago*' in relation to AI presses us to articulate a theo-ontology for dialogues between theology and AI research before tackling issues surrounding the applications of real-life AI.

0.5 The *Imago Dei as an Ontological Tool*

It is this theo-ontological objective that this study attempts to achieve based on the theology of the *imago Dei* through exploration of the following two research questions. To what extent and how can we describe AI as the image of the human being who is the *imago Dei*? To what extent can such a description help to address ethical issues related to the deployment of AI in faith communities and human society at large?

In order to address these two major research questions, this study will employ the Reformed theology of archetype–ectype as a conceptual tool to flesh out the holistic meaning of the *imago Dei*—that is, the holistic reading of the psychosomatic human being as the image of God. Archetype (ἀρχέτυπος) literally means the ultimate exemplar or pattern, and ectype (ἔκτυπος) literally refers to a copy, replica, or reflection of the ultimate pattern. In terms of theological anthropology, God is *the* archetype, and the human being is *the* ectype. This archetypal–ectypal relationship between God and humans is in conformity with the theological concept that the human being is the image of God. Since the Christian doctrine of *creatio ex nihilo* brings into light the radical ontological distinction between God as the creator and humans as creatures, it stands to reason that there is the ontological chasm between the divine archetype and the human ectype. In this light, the notion of archetype–ectype is embedded with the implication of ontological chasm, which lays the foundation for all distinction between God and humans, say, divine creation and human creative activities.

From this theological vantage point, it can be argued that AI as the artefact is the ectype of the *imago Dei* rather than the *imago Dei* itself precisely because the *imago Dei* is incapable of creating any things in the same way as divine *creatio ex nihilo*. That is to say, human beings as the *imago Dei* cannot reproduce full equivalents for the *imago Dei*. From this it follows that AI is created in the *imago hominis*. In turn, we arrive at two ontological corollaries. Firstly, given the analogical relationship between an archetype and an ectype, AI as the ectype of the human being is inextricably connected to the *imago Dei*. Secondly, by virtue of the ontological chasm between an ectype and an archetype, AI as the ectype of *the imago Dei* implies that there is an insurmountable ontological gap between humans and AI. This connected yet differentiated ontological relationship between humans and AI underpins interaction between AI and theology.

Furthermore, coupled with the theology of archetype and ectype, the *imago Dei* does not merely denote some functions, properties, or elements of humanity. Rather, the human being as psychosomatic whole *is* the *imago Dei*. Wedded to this holistic interpretation of the *imago Dei* and borrowing the findings of recent studies on theology and evolution (as will be elucidated in Chapter 1), this study will construe the *imago Dei* as the signifier of the beginning of both God–human stories and the stories of human ethical performances towards the others. Conjoining this observation and the above two ontological corollaries, I shall argue that AI's religious and ethical vistas consist in that AI as the *imago hominis* is conducive to the narration of the stories both of God and psychosomatic humans and of human ethical performances towards others. Moreover, combined with the idea of archetype–ectype, both that humans are the *imago Dei* and that AI is created in the *imago hominis* provide the foundation for the uniqueness of humankind compared to AI. I will demonstrate that this uniqueness vis-à-vis AI is manifest in liturgical activities and moral agency *in concreto*.

0.6 The Way Forward

This study aims to construct a dialogue between theology and AI through the lens of a holistic understanding of the *imago Dei* and through exploration of both religious and ethical issues surrounding AI application. To this end, I will articulate an ontological methodology for theology–AI dialogues (Chapters 1 and 2) before moving on to explore interactions between AI and Christian liturgy (Chapter 3) and between AI and theological ethics (Chapters 4 and 5).[46]

Chapter 1 will spell out an ontological basis for AI–theology studies: AI is made in the *imago hominis*, that is, the image of the human being who is the *imago Dei*. Following this, I will stress the importance of a holistic account of the *imago Dei* by drawing on recent theology-and-evolution scholarship. That is, the *imago Dei* is defined as the signifier of the beginning both of God-human stories and of stories of human ethical performances to others. Accordingly, the *imago Dei* should indicate human *conscious* response both to God and others (including the nonhuman). Furthermore, this holistic account of the *imago Dei* will be read in conjunction with the Reformed theology of archetype–ectype so as to make a distinction between the *imago Dei* and the *imago hominis*, characterising AI as the ectype of humanity, which turn out to be a theo-ontological foundation for addressing religious and ethical issues surrounding AI application.

Chapter 2 is intended to expand upon the implication of this theo-ontological foundation of AI by engaging with the debates over artificial consciousness, one of the most contentious areas in AI research. The point made in this chapter is: the thesis that AI is the ectype of humanity and made in the *imago hominis* brings into the foreground the importance of the human body—the brain in particular—for the operation of human consciousness. Given the psychosomatic unity of the human being, artificial consciousness should be understood in a simulated sense, that is, as the ectype of human consciousness. This ectypal nature of artificial consciousness underpins its distinction from yet connection with human consciousness. Moreover, artificial consciousness as ectypal demonstrate that AI as the ectype of human beings is conducive to the human *conscious* narration of both God–human stories and stories of human ethical performances towards others.

Chapters 3, 4, and 5 exemplify how AI presses forward the narration of God–human stories and stories of human ethical performances towards others through exploration of AI's role in Christian liturgy, the theological understanding of AI's moral agency, and AI as companion, provided that the ontological chasm between the human being as the archetype and AI as the ectype is sustained.

Chapter 3 probes interaction between AI and Christian liturgy, looking at recent debates over whether AI-powered artefacts can be considered religious ministers to enact liturgies. Over the past several years, AI priests have appeared in some religious activities. Arguing from the perspective of Christian theology, this study will concentrate on the Christocentric character of the Christian liturgy, which epitomises and concretises stories of God and humans. The Christocentrism of Christian liturgy turns our attention to the incarnation and consequently forces us to reckon with the human *embodiment* that is inextricably bound up with *consciously* performed liturgies. This observation entails the corollary that AI, the ectype of the human being, is overshadowed by a human minister in enacting the liturgy. For all that there is the radical difference between AI-powered and human liturgical enactment, I suggest that AI ministers are feasible under exceptional circumstances, e.g., a church during the pandemic.

Chapter 4 seeks to inquire into the moral status of AI and the idea of artificial moral agency (AMA). Many have argued that AI does have the human-level moral agency, whereas others have disapproved the idea of AMA. Aside from this either-or situation, some scholars recently suggest that AMA is partial in comparison with human moral agency (HMA). This chapter shall demonstrate that coupled with the theology of archetype–ectype, the idea of the *imago hominis* can qualify and disambiguate the term 'partial.' This partial moral agency is characterised as ectypal precisely because AI's failure to perform moral actions as psychosomatically as humans. Moreover, AMA as the ectype of HMA entails that AI's moral agency is traced back to HMA and consequently proves that the human creators, stakeholders, and users of AI are held responsible for AI's actions and impacts. In this way, the holistic reading of the *imago Dei* gives birth to a theological account of responsible AI.

Chapter 5 aims to trade on the findings of the previous chapters to address the question on how AI can be considered a human companion. It will go into details of the ideas such as artificial companion and empathy. From the vantage point of AI as the ectype of humanity, I will argue that AI is unable to be an authentic companion owing to the fact that human–AI companionship largely rules out the otherness between humans, which is characteristic of human–human companionship. This corollary does not mean that AI should be excluded from the human community. Rather, by virtue of its partial moral agency as the ectype of HMA, I will propose and expand on the idea of companionable AI. That is, AI-assisted artefacts can mediate and augment human companionship without diminishing the importance of human companionship.

These five chapters showcase how the holistic understanding of the *imago Dei*, coupled with the theology of archetype–ectype, lays an ontological foundation for theological engagement with AI while addressing religious and ethical questions related to AI application. But what are the specifics of the *imago Dei*? To this subject we are turning now.

Notes

1 Tertullian, "The Prescription against Heretics," in *Latin Christianity: Its Founder, Tertullian*, ed. Alexander Roberts, James Donaldson, and A. Cleveland Coxe, trans. Peter Holmes, Ante-Nicene Fathers, vol. III (Buffalo: Christian Literature Company, 1985), VII.

2 This sacralisation or deification of technology has been noted in the twentieth century; see, for example, Jacques Ellul, *The New Demons*, trans. C. Edward Hopkin (New York: Seabury, 1975), 47–50; Neil Postman, *Technopoly: The Surrender of Culture to Technology* (New York: Vintage, 1993), 71. However, trans/post-humanism serves as a strong case for the religious nature of technology in the twenty-first century.

3 Rory Cellan-Jones, "Stephen Hawking Warns Artificial Intelligence Could End Mankind," *BBC News* November 7, 2017, https://www.bbc.co.uk/news/technology-30290540.

4 Derek C. Schuurman, "Artificial Intelligence: Discerning a Christian Response," *Perspectives on Science and Christian Faith* 71, no. 2 (2019): 79. A classical elucidation on the ontological foundation of moral values concerning technology can be found in Hans Jonas, *The Imperative of Responsibility: In Search ofan Ethics for the Technological Age*, trans. Hans Jonas and David Herr (Chicago: University of Chicago Press, 1984), 25–50.

5 Alan M. Turing, "Computing Machinery and Intelligence," in *Parsing the Turing Test: Philosophical and Methodological Issues in the Quest for the Thinking Computer*, ed. Robert Epstein, Gary Roberts, and Grace Beber (Dordrecht: Springer, 2009), 23–66. The paper cited here and in what follows is an annotated version. Its original version was published as Alan M. Turing, "Computing Machinery and Intelligence," *Mind: A Quarterly Review of Psychology and Philosophy* 59, no. 236 (1950): 433–460.

6 Ibid., 23.

7 Ibid., 38, 39.

8 Ibid., 42.

9 Stuart Russell, *Human Compatible: Artificial Intelligence and the Problem of Control* (New York: Viking, 2019), 41.

10 Raymond Kurzweil, *The Age of Intelligence Machines* (Cambridge: MIT, 1990), 14.

11 Peter Stone et al., *Artificial Intelligence and Life in 2030: One Hundred Year Study on Artificial Intelligence 2015 Study Panel Report*, Stanford University (Stanford, September 2016), 4, https://ai100.stanford.edu/sites/g/files/sbiybj18871/files/media/file/ai100report10032016fnl_singles.pdf.
12 Margaret A. Boden, *Artificial Intelligence: A Very Short Introduction* (Oxford: Oxford University Press, 2018), 1.
13 Mark Coeckelbergh, *AI Ethics* (Cambridge: The MIT Press, 2020), 65.
14 O. Holland and D. Gamez, "Artificial Intelligence and Consciousness," in *Encyclopedia of Consciousness*, ed. W. P. Banks (Oxford: Academic Press, 2009), 40.
15 This study distinguishes AI into three categories according to its similarities or dissimilarities to human intelligence. There are other taxonomies. For example, Ashok Goel and Jim Davies divide AI into two categories: 'Engineering AI is concerned with how to design the smartest intelligent artifacts possible, regardless of whether the processes implemented reflect those found in natural intelligences. The vast majority of AI research falls into this category. Cognitive AI, in contrast, endeavors to design artefacts that think the way people (or sometimes other animals) do.' Ashok K. Goel and Jim Davies, "Artificial Intelligence," in *The Cambridge Handbook of Intelligence*, ed. Robert J. Sternberg and Scott Barry Kaufman (Cambridge: Cambridge University Press, 2011), 468.
16 John R. Searle, "Minds, Brains, and Programs," *The Behavioral and Brain Sciences* 3 (1980): 417–457.
17 AGI is defined as follows: 'The term [artificial general intelligence] is meant to stress the general-purpose nature of intelligence–meaning that intelligence is a capacity that can be applied to various (though not necessarily all possible) environments to solve problems (though not necessarily being absolutely correct or optimal).' Pei Wang and Ben Goertzel, "Introduction: What Is the Matter Here?," in *Theoretical Foundations of Artificial General Intelligence*, ed. Pei Wang and Ben Goertzel (Paris: Atlantis, 2012), 2.
18 A recent discussion on this, see Calvin Mercer and Tracy J. Trothen, *Religion and the Technological Future: An Introduction to Biohacking, Artificial Intelligence, and Transhumanism* (Cham: Palgrave Macmillan, 2021), 181–204.
19 Anne Foerst, *God in the Machine: What Robots Teach Us about Humanity and God* (New York: Plume, 2005), 67.
20 A video of Cue's shooting practice can be seen on "Meet Cue—the robot that shoots hoops," https://www.bbc.co.uk/newsround/47786237.
21 Stone et al., *Artificial Intelligence and Life in 2030*, 13.
22 Rolf Pfeifer and Fumiya Iida, "Embodied Artificial Intelligence: Trends and Challenges," in *Embodied Artificial Intelligence*, ed. Fumiya Iida et al. (Berlin: Springer, 2004), 2–4. Pfeifer and Iida narrate the history of embodied AI research from the 1980s to the early twenty-first century.
23 Foerst, *God in the Machine*, 77–111. Foerst's corollary echoes Mark Johnson's observation: human 'reality is shaped by the patterns of our bodily movement, the contours of our spatial and temporal orientation, and the forms of our interaction with objects. … [S]tructures of our bodily experience work their way up into abstract meanings and patterns of inference.' Mark Johnson, *The Body in the Mind: The Bodily Basis of Meaning, Imagination, and Reason* (Chicago: University Of Chicago Press, 1987), xix.
24 Carl Strathearn and Minhua Ma, "Modelling User Preference for Embodied Artificial Intelligence and Appearance in Realistic Humanoid Robots," *Informatics* 7, no. 3 (2020): 28.
25 Beth Singler, "An Introduction to Artificial Intelligence and Religion For the Religious Studies Scholar," *Implicit Religion* 20, no. 3 (2017): 219.
26 Martin Heidegger, *The Question Concerning Technology and Other Essays*, trans. William Lovitt (New York: Garland, 1977), 12, 39–40.

27 Justin E. Lane, *Understanding Religion through Artificial Intelligence: Bonding and Belief* (London: Bloomsbury Academic, 2021), 8; also see Justin E. Lane, “Method, Theory, and Multi-Agent Artificial Intelligence: Creating Computer Models of Complex Social Interaction,” *Journal for the Cognitive Science of Religion,* 1, no. 2 (2013): 161–180.
28 Lane, *Understanding Religion through Artificial Intelligence*, 10–13.
29 Ibid., 17, 18.
30 Ibid., 14, 21.
31 Robert M. Geraci, “Spiritual Robots: Religion and Our Scientific View of the Natural World,” *Theology and Science* 4, no. 3 (2006): 229–246.
32 Robert M. Geraci, *Apocalyptic AI: Visions of Heaven in Robotics, Artificial Intelligence, and Virtual Reality* (Oxford: Oxford University Press, 2010); Geraci’s summary discussion on Apocalyptic AI can be found in Robert M. Geraci, “Apocalyptic AI: Religion and the Promise of Artificial Intelligence,” *Journal of the American Academy of Religion* 76, no. 1 (2008): 138–166.
33 Robert M. Geraci, “Robots and the Sacred in Science and Science Fiction: Theological Implications of Artificial Intelligence,” *Zygon: Journal of Religion and Science* 42, no. 4 (2007): 966.
34 Geraci, *Apocalyptic AI*, 1.
35 Ibid., 9. Geraci argues that this apocalypticism ‘refers to (1) a dualistic view of the world, which is (2) aggravated by a sense of alienation that can be resolved only through (3) the establishment of a radically transcendent new world that abolishes the dualism and requires (4) radically purified bodies for its inhabitants’ (9).
36 Ilia Delio, *Re-Enchanting the Earth: Why AI Needs Religion* (Maryknoll, NY: Orbis Books, 2020), xvi.
37 Ibid., 76–131.
38 Ibid., 223.
39 John Calvin, *Institutes of the Christian Religion, Volume 1 & 2*, trans. Ford Lewis Battles, ed. John T. McNeill (Louisville, Kentucky: Westminster John Knox Press, 2011), 1.1.1–3.
40 Pope Francis, “The Common Good in the Digital Age,” (27 September 2019). https://www.vatican.va/content/francesco/en/speeches/2019/september/documents/papa-francesco_20190927_eradigitale.html.
41 David B. Hart and John Chryssavgis, eds., *For the Life of the World: Toward a Social Ethos of the Orthodox Church* (Brookline: Holy Cross Orthodox Press, 2020), §70.
42 William Young, “Virtual Pastor: Virtualization, AI, and Pastoral Care,” *Theology and Science* 20, no. 1 (2022): 6–22; also see William Young, “Reverend Robot: Automation and Clergy,” *Zygon: Journal of Religion and Science* 54, no. 2 (2019): 479–500.
43 Yong Sup Song, “Religious AI as an Option to the Risks of Superintelligence: A Protestant Theological Perspective,” *Theology and Science* 19, no. 1 (2021): 73.
44 Ibid., 75.
45 Eugene A. Curry, “Artificial Intelligence and Baptism: Cutting a Gordian Knot,” *Theology and Science* 20, no. 2 (2022): 156–165.
46 The summaries of several sections of Chapters 1, 4 and 5 were published to articulate a theological account of artificial moral agency; Ximian Xu, “A Theological Account of Artificial Moral Agency,” *Studies in Christian Ethics* 36, no. 3 (2023): 642–659.

1 A Holistic Reading of the *Imago Dei*

From the *Imago Dei* to the *Imago Hominis*

> We are meaning-making, conscious, embodied, and living beings whose nature, mind, and knowledge cannot be explained away by comparisons to machines.
>
> ———Mark Coeckelbergh[1]

Technological developments provide a spur on the human imagination of the other world, which is described by some scholars as a religion of technology. David Noble observes that, in the Western world, such a religion of technology is inextricably associated with Christianity.

> The other-worldly roots of the religion of technology were distinctly Christian. For Christianity alone blurred the distinction and bridged the divide between the human and the divine. Only here did salvation come to signify the restoration of mankind to its original God-likeness.[2]

Noble's claim may invite criticisms because the Christian doctrine of *creatio ex nihilo* (creation out of nothing) refrains Christianity from *blurring* the boundaries between the divine and the human and instead ensures the ontological distinction between God and creatures. Indeed, Christianity proclaims the reconciliation (or, bridging the divide) between God and humankind in Christ. Yet, reconciliation does not obliterate but rather is preconditioned by the divine–human ontological distinction. Despite Noble's misleading wording in his formulation of Christianity's stance on God–human relationship, he rightly draws our attention to how the idea of God-likeness, namely, the *imago Dei*, comes into play in the age of technology.

I have briefly argued in 'Introduction' that the *imago Dei* is a conceptual tool for an ontological configuration of artificial intelligence (AI)-and-theology studies. Having brought into the foreground the idea of AI as imaging or mimicking humans, it is imperative for theologians to dot the i's and cross the t's of the *imago Dei* while engaging with AI.

In the history of Christian theology, the *imago Dei* is conventionally construed in one of the three senses: the *imago Dei* as human rational power,

DOI: 10.4324/9781003356738-2

the *imago Dei* as human dominion over creatures, and the *imago Dei* as human relationships with others. Yet notwithstanding the conventional interpretations, recent interdisciplinary studies on theology and evolutionary theory force us to reconsider the meaning of the *imago Dei* and lead us to a holistic construal of the *imago Dei*, that is, the whole human being as the image of God.

This chapter will invoke such a revisionist approach to reconfigure a holistic meaning of the *imago Dei*. That is, the *imago Dei* is the signifier of the beginning of both God–human stories and stories of human ethical performances towards others. This holistic account entails that we need to take the *whole* human being into consideration while engaging theologically with AI, exploring both the human mental and physical factors. Teamed up with the Reformed theology of archetype–ectype, this holistic construal elucidates that being created in the *imago Dei*, the human being is the ectype of God who is the archetype. Hence, there is the ontological chasm between the divine archetype and the human ectype, which, in turn, gives rise to the radical distinction between divine and human creative activities. As an artefact, AI cannot be considered the *imago Dei* but rather the *imago hominis* and the ectype of humanity precisely because the *imago Dei* is the consequence of God's creation, that is, *creatio ex nihilo*. This distinction between the *imago Dei* and the *imago hominis* brings forth the theo-ontology of AI, which—as the above epigraph indicates—demonstrates the uniqueness of the human being vis-à-vis AI.

In the following text, I will first sketch out how the idea of the *imago Dei* was employed to facilitate theology-and-AI dialogues in previous studies, examining Noreen Herzfeld's ground-breaking work on AI and theology at close quarters. From this it follows that theology-and-AI studies, by and large, fall short of a holistic understanding of the *imago Dei*; neither do they capitalise on this conceptual tool to address ethical issues related to AI. Second, I will turn to recent holistic interpretations of the *imago Dei* developed through theological engagement with evolutionary theory and with particular attention to David Fergusson's and Celia Deane-Drummond's studies. Third, I will proceed to spell out the Reformed notion of archetype–ectype and clarify how this theological concept can fortify the idea of the *imago Dei* and construct a theo-ontology of AI. Finally, I will flesh out the idea of AI as the ectype of humanity and as the *imago hominis*, followed by illuminating its ontological importance for dealing with religious and ethical issues related to AI.

1.1 AI and the *Imago Dei*

The idea of the *imago Dei* has been featured in interdisciplinary studies on theology and AI for many years. A number of attempts were made to use the *imago Dei* as a conceptual tool through which theological responses to challenges and issues surrounding AI could be articulated.

Noreen Herzfeld offers the first extensive exploration of AI from the perspective of the *imago Dei*. She aims to deal with the major research question: 'What, precisely, are the qualities or capacities that we hold as so important to our human nature that we hope to image them in AI, and how might our choice among these qualities affect any intelligent machine we produce, our relationship with that machine, and our understanding of ourselves?'[3] This question exhibits Herzfeld's conviction that what is imaged in AI reflects what is essential and central to humans, and that humans attempt to project their nature onto AI. (This conviction has been cogently illustrated in 'Introduction'.) It is precisely based on this contention that she makes use of the idea of the *imago Dei* as a point of contact between theology and AI. The *imago Dei* serves as a conceptual apparatus to investigate the mutual influence between theological anthropology and AI research.

Herzfeld's study is structured according to the three conventional interpretations of the *imago Dei*: the substantive, the functional, and the relational.[4] As will be seen, following Herzfeld, many scholars pay their attention to one of the three interpretations to explore the interaction between Christian theology and AI research. To this extent, Herzfeld's study affords a framework into which other scholarly discussions on AI and the *imago Dei* can be set.

Herzfeld first examines the substantive approach to the understanding of the *imago Dei* and its consequent effect on our understanding of AI. The substantive approach identifies human reason as the essence of the *imago Dei*.[5] Human image-bearing primarily denotes the rational activities of the human mind, and the *imago Dei*, by its nature, amounts to human rationality. Herzfeld points out that this substantive interpretation resonates with symbolic AI (weak AI), which assumes 'a finite set of elementary symbols' that are combined 'into more complex patterns, ideas, or thoughts' according to 'finite sets of rules.'[6] Symbolic AI is designed to replicate human intelligence and act in place of humans to perform some intellectual activities in limited domains.

Herzfeld contests that if AI is understood in substantive sense, the human body is *de facto* denigrated and its importance for human intelligence is undermined.[7] Her observation strikes a chord with many scholars, especially those who are supportive of embodied AI research. As we noted in 'Introduction', embodied AI approach emphasises the embodiment of AI system as something that is of importance for an *intelligent* agent. On this front, AI research is in tune with recent scholarship on theological anthropology, which lays emphasis on the human body. For example, Nancey Murphy conjoins theological analysis and scientific findings to show the perpetual interaction between the body and the soul.[8] Viewed in this light, a dualistic view on human psychic and physical dimensions is, both theologically and scientifically, untenable. The substantive interpretation of the *imago Dei* cannot gain ground in scientific or theological fields and, as such, fails to theologically exhibit the *imago Dei* as a point of contact between AI and theology. We will discuss the importance of human body for the idea of intelligence time and again throughout the rest of this study.

The second interpretative method of the *imago Dei*, coming under Herzfeld's scrutiny, is functional, which characterises the *imago* as regency. That is, the human being acts as God's deputy to exercise dominion on earth.[9] This interpretation ties in with the view of weak AI. Rather than seeking to build human-level intelligence, weak AI is content with replicating certain functions of the human mind in order to solve problems and complete tasks in limited domains. Like the human being as God's deputy, weak AI is nothing other than a tool or servant of humankind for completing specific tasks.[10]

Herzfeld points out two drawbacks of weak AI from technological perspectives. First, the boundary between weak AI and a normal computer application is not clear. Both weak AI and a computer application are programmed to perform certain tasks. If AI is considered in a function-oriented sense, the boundaries between weak AI and computer applications become blurry because some tasks completed by weak AI can also be achieved by some computer applications. Second, 'once an application is mastered, there is a tendency to no longer define that application as AI' because those tasks are not completed in the same way as a human person would do.[11] In this light, the functional interpretation may result in making weak AI and computer applications identical.

Moreover, Herzfeld disputes the possibility of creating strong AI according to the functional interpretation of the *imago Dei*. For it is difficult to define which function constitutes intelligence. On the other hand, unlike humans who can perform functions harmoniously, AI programmes are designed to complete specific tasks and are thus seldom congruent with one another.[12] In addition, the functional interpretation raises ethical concerns. The salient issue is that AI simply becomes the instrument with which humans expand their dominion over the world, which would challenge the human existence itself. For example, humans would be induced to act and work in an increasingly mechanical way, which might lead humans to be identified as machines.[13] In short, Herzfeld takes issue with the functional interpretation of the *imago Dei* and its consequent effect on our view of (weak) AI.

Recent studies on theology and AI pay close heed to the functional approach. In 'Performing the *Imago Dei*,' Karen O'Donnell crafts a performative route to developing the doctrine of the *imago Dei* as a theological foundation for AI. Her main argument is constructed upon Alistair McFadyen's performative interpretation of the *imago Dei*:

> The primary task of theological anthropology is not to give an account of universal human nature, nor to provide supposedly universal definitions of the image of God. Its task is not to define the image of God but to image God; to be commentary on and participant in God's active seeking of humanity in its full flourishing, focused on those places where humanity and its flourishing are counterfactual. This makes the image more like a verb than it is a noun; more dynamic than it is static; more performative than indicative.[14]

In McFadyen's position, the *imago Dei* does not simply connote a static status. Rather, it indicates a dynamic action—imaging God, which 'represents our being caught up in the active presence and movement of God towards the world creatively, redemptively and eschatologically.'[15] Based on McFadyen's idea of imaging God, O'Donnell goes on to illustrate that AI is the *imago Dei* rather than the *imago hominis*.

It is worth pausing here to clarify that O'Donnell deliberately distinguishes her performative approach from the functional reading of the *imago Dei*. She asserts that 'a performative understanding of the *imago Dei* differs from the functional interpretation in that it is not based on regency or dominion, but rather on orientation and action and the seeking of the image of God in the other.'[16] However, considering from the task-oriented perspective, the performative interpretation of the *imago Dei* is on a par with the functional interpretation. The performative approach is intended to construe the *task* and *function* of imaging God's action and movement towards creatures. The performative interpretation of the *imago Dei* is the disguised functional interpretation.

O'Donnell suggests that there are three performances of the *imago Dei* with reference to technology: the performances of creation, hope, and wholeness. The latter two performances refer respectively to 'the hope of a restored life and ability to engage fully with creation' and the integration of technology with the human body as a whole to enhance humanity.[17] AI is pertinent to the first performance of the *imago Dei*. With the creation of AI, the 'imaging of God happens as this performance of creativity is brought into dialogue with the active seeking of the image in the other. It is in the performance of creativity and creation that one images God and humanity is found.'[18] Following this, O'Donnell envisages the scenario of strong AI in reference to the performances of hope and wholeness. She is convinced that if AI 'is autonomous and can learn, as a new Christian does, to perform the image of God and seek it in the other in specific, concrete situations,' it can be considered the *imago Dei* by virtue of its performance of the *imago*.[19]

O'Donnell gets around several challenges—as raised by Herzfeld—to the functional approach to the *imago Dei*. For O'Donnell, AI is not an instrument at all; neither is it used to expand the human dominion over the world. It is the other way around that AI becomes the companion of humans by performing the *imago Dei*, which ties in with the relational interpretation of the *imago Dei*, as will be explicated anon. However, I am less convinced than O'Donnell that the *imago Dei* can be forthrightly employed to qualify AI on the grounds that O'Donnell's argument rests in the presupposition that a human-level consciousness can be created within the AI. For O'Donnell, a strong AI-powered agent can *consciously* perform the *imago*, namely, the imaging of God's movement towards the world. However, as will be illustrated, the theology of archetype–ectype underwrites that the idea of the *imago Dei* implies an ontological gap between humans and AI.

Having criticised the substantive and functional interpretations, Herzfeld critically appropriates the relational interpretation of the *imago Dei* to construct a dialogue between theology and AI.[20] The relational interpretation starts from the doctrine of the Trinity, arguing that the intra-Trinitarian relationship and communion lay the foundation for all relationships *ad extra*.[21] Herzfeld's sympathy with the relational interpretation rests in her conviction that the 'ability to converse is not just a test of intelligence. It may also be a precondition.'[22] Since conversations presuppose relationships between agents, 'the idea of an individual intelligence is meaningless; intelligence has meaning only in encounter.' AI is not an individual agent but belongs to the human community.[23]

The relational reading of the *imago Dei* comes to the point that AI is able to relate to humans through conversation and becomes the companion of humans. Herzfeld suggests that the human–AI relationship is under the rubric of human–nonhuman relationships and conveys moral significance, which shows our responsibility for the natural world.[24] Despite her sympathy with the relational interpretation of the *imago Dei*, she issues two major warnings. First, the relational approach may result in the precedence of AI–human relationships over human–human relationships, which are situated in the context of the God–human relationship. Second, a graver outcome is idolatry.[25]

> This is idolatry, for when we project our own attributes and desires onto God we create a god who is our own fabrication, embodying our values and satisfying our narcissistic whims. There is a similar risk in the image connecting humans to computers. As we give computers more and more human traits we have a tendency to see them not as our tools but as our children. But we are infinitely more adaptable than our machines. Viewing our machines as fellow creatures is unlikely to change them, and more likely to subtly shape our image of ourselves, of what our lives and relationships are about. This, too, is a form of idolatry, for insofar as we allow computers to shape our lives we have made them into gods of silicon.[26]

Accordingly, the stand or fall of the relational approach to AI depends upon whether or not the God–human relationship is always prioritised. Although this study does not follow the above three conventional interpretations of the *imago Dei*, it will demonstrate that Herzfeld's warning is in line with one rationale of my methodology—that is, the God–human relationship is always dominant on the human–AI relationship.

In distinction to Herzfeld, Chammah Judex Kaunda makes a more optimistic claim that the relational approach to AI is promising. Kaunda develops his argument by criticising Herzfeld's contention that humans 'must always be aware of the otherness of any artificial intelligence.'[27] On the contrary, Kaunda contends that 'Herzfeld's "othernization" could be taken to imply only human[s] are the legitimate *imago Dei*, other creation only shares on human terms. She failed to explain the ontological relation of AI to humanity. Thus, socioreligious relations and *imago Dei* remain human possessions.'[28] Following this, he argues that 'God is the only original creator, humans are created creators, and that every aspect of visible existence, including AI, is inherently divine imaging.'[29] In order to support his argument, Kaunda deploys the Bemba idea of mystic relationality to demonstrate the possibility of strong AI's participation in the *imago Dei*. 'Mystico-relationality ... refers to the mystical solidarity that all creation shares because of their participation in the divine life.'[30] From this it follows that the physical and the spiritual are never independent of each other; instead, they are in continuity.[31]

Bearing in mind this fundamental rationale, Kaunda goes on to submit that humans as created in the *imago Dei* are called to participate in divine creativity in such a way that the human person as the created co-creator serves as a secondary agent to actualise God's continuous creation.[32] This positive position towards human creative activities underpins his endorsement of AGI or strong AI, which is capable of participating in the image of God: 'by virtue of being an aspect of primal creation of God, as an aspect of God's material world, AGI like all creation participates ... "weak imaging" of God. The weak imaging is intrinsic to all God's creation that allows them to generally participate in God's life.'[33] That being so, there is the ontological connection between humans and AGI in virtue of the fact that both are created by God and thus participate in the primal image of God.[34] In Kaunda's estimation, AGI's participation in the *imago Dei* enriches and broadens the *imaging* of God on all levels of God's creation.[35]

Through the above analysis, it comes to be seen that Kaunda is free of Herzfeld's caveats by emphasising the conjunction of human and AI's participation in the *imago Dei*. Although Kaunda's attention to the ontological connection between AI and humans is commendable, and such a connection will be unfolded in this study, I take issue with his position because there is a subtle logical leap in his argumentation. Kaunda is so hasty to form the coalition between humans and AI that he is oblivious to the fact that AI is designed by mimicking humans, namely, being made in the *imago hominis*. To put it differently, Kaunda takes a logical leap to obliterate the difference between the *imago Dei* and the image of the *imago Dei*. This study is intended to refrain from such a logical leap and retain the ontological connection *yet distinction* between the *imago hominis* and the *imago Dei*.

In summary, the dialogues between Christian theology and AI constructed with the *imago Dei* as a conceptual tool bring to light many relevant religious and ethical issues. Herzfeld's exploration of the three approaches to the interaction between the *imago Dei* and AI provides a framework in which other studies can be categorised. Moreover, each approach shows certain limitations. The substantive approach devalues the importance of the human body for cognition. The functional approach elicits technical difficulties and ethical concerns about the expansion of human dominion. Although the relational approach is, according to Herzfeld, the best of the three, it nevertheless may cause chaos among human–human and God–human relationships and blur the distinction between the *imago hominis* and the *imago Dei*. Yet notwithstanding the shortcomings of the three conventional interpretations, this study still maintains the stance that the *imago Dei* remains to be a strong link between theology and AI insofar as the idea of the *imago Dei* can give birth to a theo-ontology for AI, which conduces to addressing both religious and ethical issues related to AI. This theo-ontology can only be articulated through a holistic re-interpretation of the *imago Dei*.

1.2 A Holistic Account of the *Imago Dei*

In his Gifford Lectures, J. Wentzel van Huyssteen points us to a contemporary challenge to the idea of the *imago Dei*:

> The biblical idea that human beings are created in the image of God and that *Homo sapiens* are ... placed more or less at the center of the created universe seems to be especially challenged in our time. For Christian theology one of the most crucial questions today should be whether there is a way in which we may rediscover the canonical function and orienting power of a concept like the *imago Dei* without retreating to metaphysical abstractions.[36]

David Fergusson's and Celia Deane-Drummond's interpretations of the *imago Dei* offer responses to this challenge. They do not retreat to metaphysics but rather stress that other nonhuman life forms need to be considered while construing the *imago Dei*. Their emphases are slightly different. While Fergusson aims to construct a holistic account of the *imago Dei* theologically so as to articulate the *imago Dei* in view of evolution, Deane-Drummond attends more to the ethical relationship between the *imago Dei* and animals through the lens of theology-and-evolution dialogues.

Fergusson's interpretation of the *imago Dei* underlines human particularity in the context of evolution: 'What evolutionary history reinforces ... is the extent to which our particularity is embodied and embedded in a wider narrative about creaturely existence. And the account we offer of that particularity will reveal strong links to other life-forms.'[37] His account of human particularity in an evolutionary world stems from his holistic account of the *imago Dei*. In 'Humans Created according to the *Imago Dei*,' Fergusson makes clear his purpose to offer 'a more holistic description that includes functional, relational, and practical elements' apart from the conventional approaches.[38] That is to say, rather than overturning the three conventional interpretations of the *imago Dei*, he seeks to articulate a framework into which the substantive, the functional, and the relational understandings can be fit. In his position, although the idea of the *imago Dei* appears in the first chapters of the Bible, 'its meaning is opaque.'[39] It is implied in this observation that the three conventional interpretations of the *imago Dei* are likely to read a-biblical elements into scriptural narratives. As a result, Fergusson, first of all, pinpoints the drawbacks of the three conventional interpretations of the *imago Dei* before establishing the holistic framework.

Fergusson maintains that what is at issue in the substantive interpretation of the *imago Dei* is the ontological privilege of humankind, which is at variance with scientific findings that show 'a much greater continuity of human beings with other species.' Moreover, this interpretation deviates from the holistic Hebrew anthropology and recent exegeses of Genesis 1:26-27; the latter casts it into doubt that the *imago Dei* can 'be identified with a distinctive noetic feature or single component of human existence.' The Christological and soteriological implications of the *imago Dei* in the New Testament need to be taken into consideration. At the same time, Fergusson observes that the substantive interpretation would result in ecological problems, belittle the 'moral status of other animals,' and give rise to biases among human beings, viewing 'some human beings as more exemplary of the divine nature than others.'[40]

Fergusson contests that although the relational interpretation of the *imago Dei* eludes the above issues incurred by the substantive interpretation, it nevertheless brings about two problems. First, the relational interpretation reads the intra-Trinitarian relationship into Genesis 1:26–27; these two verses themselves do not necessarily lead to such a Trinitarian exegesis. Second, the relational interpretation is hasty to move analogically from the Trinity to creatures. 'The primary intention of the doctrine of the Trinity was never anthropological, as if the divine society were the archetype of a model human community.'[41] To put it differently, the relational interpretation runs the risk of surrendering the doctrine of the Trinity to theological anthropology.

Fergusson moves on to examine the functional interpretation, arguing that it is 'an excessive anthropocentrism,' though it is more in line with the original context of Genesis passages than the other two interpretations.[42] This functional interpretation often develops with an emphasis on 'pastoral care and responsible stewardship.' Nonetheless, Fergusson argues that this stewardship remains to be anthropocentric.[43] Moreover, he notes that Genesis passages do not identify which functions constitute the *imago Dei*.[44]

Alternative to the three conventional interpretations, Fergusson furnishes an elaborate account of the holistic idea of the *imago Dei*:

> the *imago Dei* might be better understood as a marker or signifier of a history that will unfold ... What it signifies is that human beings have a central and distinctive place in the story that is about to be told. ... The story that is about to be narrated is the story of human beings and God, although other actors are also present ... While the *imago Dei* signifies here our identity as God's covenant partners, what this entails will require subsequent narration in the stories of Noah, Abraham et al.[45]

The holism of Fergusson's definition of the *imago Dei* is made up of two aspects. First, humankind as a whole is the *imago Dei*. All humans are God's covenant partners and participate in the story of God and humans. Second, the human being as a whole, including both the psychic and physical dimensions, belongs to the *imago Dei* and involves in the story of God and humans.

This holistic account also brings into the foreground the religious implication of the *imago Dei*. That is, the meaning of the *imago Dei* lies in the encounter between God and humans. Furthermore, by making reference to Celia Deane-Drummond, Fergusson observes the ethical implication of this holistic account of the *imago Dei*; that is, it refrains from denigrating nonhuman creatures because they 'take their appointed place' in the unfolding of God–human stories.[46] As such, this holistic account of the *imago Dei* offers a new lens through which the *imago Dei* can be used to construe human particularity in an evolutionary world.

Fergusson's holistic definition resonates with Celia Deane-Drummond's reading of the *imago Dei* in the sense of performing. Deane-Drummond maintains that theological anthropology should reckon with 'the becoming and being of the human through being and act, rather than simply in either ontological or functional or relational terms.'[47] Like Fergusson, she aims to use the performative understanding of the *imago Dei* to '[incorporate] rather than [reject] more traditional constitutional elements of reason and freedom, for example.'[48] She submits that inasmuch as the term '*imago Dei*' is often used in conceptual relation to anthropocentrism, 'image-bearing' carries a more profound meaning. It refers to 'a performance [that is] caught up in a shared drama with other species and orientated toward God's purposes for history.'[49] In distinction to Fergusson, she is more attentive to the ethical significance of the *imago Dei*, which means human ethical performances towards others.[50]

From the perspective of evolutionary theory, Deane-Drummond delineates the ethical implication of human image-bearing in relation to other creatures, particularly to animals.

> Human beings have evolved in such a way that their lives are entangled with those of other animals, and morality grew out of close relationships between kin in a wider multispecies context. ... If the way we love and the way we express our freedom has to imitate God in order to express the divine image.[51]

It is evident that Deane-Drummond spells out the *imago Dei* in a dynamic way. Lying at the gravity centre of her performative understanding of the *imago Dei* is the human active performance towards other creatures in imitating God.[52] In this way, she turns down attempts made to simplify the *imago Dei* to any element of the being of humans. All the more important is that she underscores the correspondence between this performative understanding and the Christological implication of the *imago Dei*:

> in order to move from image-bearing as that which is characteristic of all human beings by reason of being human to one that is a more active reflection on what such an image enables, the possibility of a transformative movement toward the image of the Son becomes the goal for the Christian life.[53]

This Christological dimension of the *imago Dei* points us to the imitation of Christ's humble service, which, in turn, highlights our ethical responsibility to 'make room for the creatures who share our creaturely home.'[54]

It is evident that behind Deane-Drummond's focus on human ethical responsibility towards other animals lies her belief in the religious aspect of the *imago Dei*. Inasmuch as the *imago Dei* carries the connotation of the human duty towards God, human image-bearing should exhibit the human duty towards other creatures. The close tie between the religious and ethical dimensions of the *imago Dei* is predicated upon the human religious capacity that evolved.[55] For Deane-Drummond, the history of human evolution shows that, from the beginning of creation, the *imago Dei* is embedded with religious and ethical implications, which are closely connected to other evolving creatures.

Having examined the issues caused by the three conventional interpretations, Fergusson and Deane-Drummond articulate the more holistic understandings of the *imago Dei*, impelling us to go beyond the conventional interpretations. Their revisionist understandings of the *imago Dei* sit well with the contemporary evolutionary idea of human niche. Agustín Fuentes argues, 'The human niche is the spatial, ecological, and social sphere that includes all social partners, perceptual contexts, and ecologies of human individuals, groups, and communities and the many other species that live with and alongside humans.'[56] With the idea of niche in mind, Deane-Drummond notes that '[i]f, in the light of niche construction theory, we shift to a much more organic view of human creativity and response, then the natural environment is not some blank stage, but itself becomes woven into the fabric of the drama into which human lives are situated.'[57] The idea of human niche strongly indicates that both human individuals and the human community interact with circumstances and other species. Undoubtedly, Fergusson's and Deane-Drummond's revisionist accounts of the *imago Dei* should provide spurs on the dialogues between theology and AI insofar as AI has become a significant part of the human niche.

Bringing together Fergusson's and Deane-Drummond's viewpoints, a holistic account of the *imago Dei* can be developed to undergird Christian theology's engagement with AI research. That is, the *imago Dei* is the signifier of the beginning of both the God–human story and the stories of the human person's performances towards others, be it human or nonhuman. This holistic account can advance the interplay between theology and AI in two ways. First, it underlines the holism of the human being, emphasising both the soul and the body, and avoids attributing the *imago Dei* to the human soul or psychic capacities, such as rationality, as raised in the substantive interpretation. Rather, it offers an integrated theological anthropology, which can be used to deal with the issues raised in AGI and ASI. As will be seen, the proponents of AGI and ASI, who endorse the possibility of a human-level consciousness in AI, often overlook the biological particularities of human beings. Second, this holistic account of the *imago Dei* makes it clear that the meaning of being human is partly predicated upon *others*. God–human, human–human, and human–nonhuman stories together make up the meaning and value of being human. This is reminiscent of David Kelsey's view of the eccentricity of human beings: 'the reality and value of human beings, their basic personal identities, and how they ought to be set into and oriented toward their ultimate and proximate contexts are all *eccentric, grounded outside themselves* in the concrete ways in which the triune God relates to all that is not God, including humankind.'[58] Accordingly, this eccentricity lies at the centre of being human, which is to say, being the *imago Dei*. From this it follows that this eccentricity of human beings plays a role in the interaction between theology and AI through the theological apparatus of the *imago Dei*. That is, the holistic account of the *imago Dei* indicates that the human–AI relationship is part of the meaning and value of being human. (The implication of this observation will become evident while exploring the extent to which AI is considered the human companion in Chapter 5.)

I will return to these two aspects time and again throughout this study. Before expanding on this holistic notion of the *imago Dei* in relation to AI, we need to address the following questions: How can the holistic view of the *imago Dei* be used to facilitate theology-and-AI dialogues? Is AI created in the *imago Dei*? If not, how can we use the holistic view of the *imago Dei* to construct a theo-ontology for AI and address theological and moral issues surrounding AI applications?

1.3 The Reformed Theory of Archetype–Ectype

As illustrated in the Section 1.1, Noreen Herzfeld and other scholars have endeavoured to sketch out the dialogues between theology and AI through the conceptual apparatus of the *imago Dei*. Indeed, the fact that AI is designed by mimicking human beings justifies using the notion of the *imago Dei* in such dialogues. However, it cannot be taken for granted that the *imago Dei* by itself can be related to AI insofar as the *imago Dei* is a term used to elucidate theologically what the human being is. Hence, before speaking of AI in relation to the *imago Dei*, we need to search for an ontic connection between them, asking whether and how AI can be ontologically linked to AI. I suggest that the Reformed theology of archetype–ectype can theologically tailor the notion of the *imago Dei* to this ontic connection.

1.3.1 Reformed Notions of Archetype and Ectype

The notions of archetype and ectype are not esoteric in the Reformed as well as Christian tradition. On the contrary, this pair of theological notions has been featuring in Christian theology since the Medieval age. Nevertheless, it was in the Reformed tradition that this conceptual pair was fully integrated into dogmatic theology. The Reformed ideas of archetypal theology and ectypal theology appeared in tandem with the rise of Reformed prolegomena.[59] Reformed theologians used the idea of archetype–ectype to describe what theology is. Yet, it is worth noting that at the basis of this idea lies a theological ontology that rests in the doctrine of creation. I now proceed to unpack the notions of 'archetype' and 'ectype' so as to demonstrate how the idea of archetype–ectype serves to connect the *imago Dei* to AI ontologically.

Franciscus Junius (1545–1602), who studied in Geneva with John Calvin (1509–1564), was the first Protestant theologian to make the distinction between archetypal (*theologia archetypa*) and ectypal theology (*theologia ectypa*).[60] He defines archetypal theology as follows: 'Archetypal theology is the divine wisdom of divine matters. Indeed, we stand in awe before this and do not seek to trace it out.'[61] Following this definition, he lays out several features of archetypal theology.

First, the word 'wisdom' does not mean the genus of archetypal theology. Based on God's simplicity, the term 'genus' cannot be attributed to archetypal theology insofar as 'genus' implies distinctions, which cannot be predicated of God because 'whatever exists in God is God.'[62] In God Himself, being and knowledge, ontology and epistemology, belong together. Second, Junius adds that 'this wisdom is divine so that we might explain that we are here treating the essential and preeminent wisdom of God which is borne above and beyond every genus.'[63] As such, this divine wisdom essentially differs from all human wisdom and gives birth to the principles of human wisdom.[64] Third, having stressed the distinction between the divine and the human, Junius describes archetypal theology as incommunicable, uncreated, essential, absolute and most perfect, and infinite.[65] By arguing so, he intends to emphasise the correlation between archetypal theology and the essence of God: archetypal theology 'is not present in God by parts nor successively, nor does it vary in itself, but exists in proportion to God Himself. It exists with His essence simultaneously, indivisible and immutable.'[66]

It is worth noting that Junius attempts not to sever human knowledge of God from archetypal theology thoroughly. He suggests that the incommunicable archetypal theology can be communicated to humans, 'not of [God's] own being properly, but precisely of His own image the most.'[67] The communication of the incommunicable archetypal theology generates ectypal theology: 'Ectypal theology, whether taken in itself, as they say, or relatively in relation to something else, is the wisdom of divine matters, fashioned by God from the archetype of Himself, through the communication of grace for His own glory.'[68] A critical point made in Junius's definition of ectypal theology is that this theology stems from, and is the reflection of, archetypal theology. By their very natures, archetypal and ectypal theologies differ from each other. In Junius' position, the fundamental distinction between archetypal and ectypal theologies rests with the theology of creation:

> For this one [ectypal theology] is created, it is dispositional; nor is it absolute except in its own mode, but rather finite, discrete, and divinely communicated. It is, as it were, a true and definite image of that theology [archetypal theology] which we have explained is uncreated, essential or formal, most absolute, infinite, at once complete, and incommunicable.[69]

By this contrast, Junius brings into explicit a crucial rationale that underlies the distinction between archetypal and ectypal theologies, that is, the ontological distinction between the created and the uncreated.

Junius's idea of archetype–ectype, along with this rationale, was formative to Protestant theology. Following Junius, Protestant orthodox theologians, including both Lutheran and Reformed theologians, took note of the idea of archetype–ectype while developing their own theology.[70] By the first quarter of the seventeenth century, this idea became common in Protestant theological work.[71]

The idea of archetypal–ectypal occupies a significant place in *Synopsis Purioris Theologiae* (1625), the acclaimed handbook of scholastic Reformed theology. In the first disputation 'Concerning the Most Sacred Theology,' the distinction between archetypal and ectypal theologies is brought into the foreground:

> Theology is of one kind if we consider the matter in itself; it is multiform if we consider the way in which the matter consists in its various subjects. For with respect to God, insofar as Theology is the knowledge whereby He knows himself and all divine things in his divine way, it is archetypal theology. ... And if Theology is viewed insofar as it is the knowledge that God either has communicated to created beings endowed with understanding in this age, or that He will share in the age to come, it is ectypal theology.[72]

It is explicit that the distinction between creatures and God as the Creator underlies the idea of archetype–ectype in *Synopsis Purioris Theologiae*. The knowledge of God communicated to creatures is always ectypal, and creatures can never get hold of God's self-knowledge or the knowledge of God in Himself.

When it came to the second half of the seventeenth century, Junius's idea of archetype–ectype is almost replicated in Francis Turretin's (1623–1687) *Institutio Theologiae Elencticae* (1679–1685), one of the greatest works on Reformed dogmatics in Reformed traditions.[73] Turretin maintains that true theology is made up of archetypal and ectypal theology:

> True theology is divided into: (1) infinite and uncreated, which is God's essential knowledge of himself (Mt. 11:27) in which he alone is at the same time the object known (*epistēton*), the knowledge (*epistēmōn*), and the knower (*epistēmē*), and that which he decreed to reveal to us concerning himself which is commonly called archetypal; and (2) finite and created, which is the image and ectype (*ektypon*) of the infinite and archetypal (*prōtotypou*).[74]

To be sure, there are overlaps in Junius's and Turretin's nomenclature used to construe the idea of archetype–ectype. Furthermore, as in Junius's thought, the rationale of the difference between the created and the uncreated underpins Turretin's distinction between archetypal and ectypal theologies.

To sum up, the Reformed idea of archetype–ectype is tied up with both the doctrine of God and the doctrine of creation: God as the creator wills to communicate the knowledge of God to creatures in an ectypal way. Derived from this claim is the ontological chasm between the archetype and the ectype. While the ectypal is always concerning what is created, the archetypal *per se* has nothing to do with the created. This ontological chasm—underpinned by the doctrine of creation—can never be overcome. In the following sections, I will spell out how the idea of archetype–ectype is deployed in theological anthropology and consequently opens up the possibility to establish an ontic connection between the *imago Dei* and AI.

1.3.2 The Notion of Archetype–Ectype in Theological Anthropology

In the Reformed tradition, the notion of archetype–ectype is often used to draw the distinction between God's self-knowledge and the knowledge of God given to humans. Few theologians deploy this notion into theological anthropology, much less into the construal of the *imago Dei*. Nonetheless, Francis Turretin offers a concise yet stimulating elaboration on the notion of archetype–ectype in relation to the *imago Dei*.

> [I]mage signifies either the archetype (*archetypon*) itself (after whose copy something is made) or the things themselves in God (in the likeness of which man was made); or the ectype itself, which is made after the copy of another thing, or the similitude itself (which is in man and the relation to God himself). In the former sense, man is said to have been made in the image of God; in the latter, however, the very image of God.[75]

Turretin sheds light on two implications of the *imago Dei* in reference to the notion of archetype–ectype. In terms of archetype, the word 'image' indicates that the human existence depends upon God. In the sense of ectype, the human being *is* the *imago Dei*. Turretin correlates the *imago Dei* with the notion of archetype–ectype so as to articulate an ontological distinction yet connection between God and human beings.

Turretin's argument remains to be developed in order to deal with some expected questions.[76] One of these questions, which is related to this study, is: how is this archetypal–ectypal understanding of the *imago Dei* relevant to the being of humans? In turn, the answer to this question will force us to consider whether the being of humans underpinned by the notion of archetype–ectype is mirrored or programmed in AI. The turn-of-the-century Dutch theologian Herman Bavinck (1854–1921) extensively deploys the notion of archetype–ectype into his theological anthropology.[77] By bringing together both a holistic account of the *imago Dei* and the notion of archetype–ectype,

Bavinck's theological anthropology can open up a way to deal with the ontological and ethical issues raised in dialogues between theology and AI.

Bavinck's theological anthropology begins with exploration of the essence of humans. His opening remark is: 'The essence of the human being lies in that he is the image of God.'[78] By arguing so, he stresses the two aspects of his holistic account of the *imago Dei*: (1) the whole human being is ontologically the *imago Dei*; (2) the *imago Dei* should be understood collectively.

Bavinck maintains that the human being *is*, in essence, the *imago Dei* rather than 'hav[ing] or bear[ing] the *imago Dei*.'[79] The ontological implication carried by the word 'is' consists in the notion of archetype–ectype, which Bavinck employs to clarify the claim that 'humans are created in God's image and likeness' (Genesis 1:26–27; 5:3; 9:6). For all that the terms 'image' and 'likeness' are used interchangeably, he observes a nuanced difference between these two terms, which is predicated on the notion of archetype–ectype:

> The likeness is a further specification, an intensification and complement of the image. 'Likeness' in itself is weaker and broader than 'image;' an animal has some features that are similar and common with humans, but is still not the image of the human being. 'Image' expresses that God is the *archetype* and the human being is the *ectype*; 'likeness' adds that this image corresponds in all parts to the original.[80]

Accordingly, the terms 'image' and 'likeness' together underscore that the human being as a whole is the *imago Dei*.[81] He explains that the terms 'image' and 'likeness' are not used to express something in God, who is 'the *imago increata* or archetype,' but rather indicate something in the human being, who is 'the *imago creata* or ectype.'[82] Bavinck makes it clear that humans are never on a par with God due to the ontological chasm between the Creator and creatures, which is underlined in the traditional Reformed notion of archetype–ectype. By the same token, this ontological chasm underlies the idea of the *imago Dei*, and so the *imago Dei* in itself implies the ontological distinction between the one who images and the one who is imaged.

Following this archetypal–ectypal understanding, Bavinck sets forth the five factors that constitute the *imago Dei*.[83] First, the *imago Dei* means that God is more demonstrable in the soul than in the body because 'the living soul is the essence of the human being.' Second, human psychic faculties belong to the *imago Dei*. Third, an essential part of the *imago Dei* is human morality, including the knowledge of moral law, righteousness, and holiness. Fourth, the *imago Dei* refers to a psychosomatic union of the body and the soul and thus incorporates the human embodiment. In this respect, Bavinck gets around the issue caused by the substantive reading of the *imago Dei*, which separates the physical from the psychic. Fifth, the *imago Dei* indicates the destiny of humankind, pointing eschatologically to 'human dwelling in paradise.' These five factors together show the core meaning of the *imago Dei*, that is, the human being as the ectype of God who is the archetype.

Having elucidated the five implications of the *imago Dei*, Bavinck conveys a holistic understanding of being human: *the whole human being, including the physical and the psychic, is the imago Dei and the ectype of the divine archetype*. Moreover, the human being as the *imago Dei* is both religious and ethical because the human being is the ectype of God and has moral knowledge. In Bavinck's theological system, both religion and ethics are bound up with human consciousness. In terms of religion, the human consciousness of God and the feeling of dependence is essential to religion.[84] At the same time, the human moral life is connected to the moral consciousness, and consequently religion and ethics are bound together in human consciousness.[85] In addition, Bavinck forthrightly correlates religious and moral consciousness with the *imago Dei*. He writes that 'the human person is the image and offspring of God, ... [and] he possesses a capacity in the consciousness (νουϛ) to see God in His works and has the work of the [moral] law written in his heart.'[86] With this in mind, the claim that the whole human person is the *imago Dei* means that the human consciousness can operate in both religious and moral fields to demonstrate that the human being is the ectype of God. The idea of religious and moral consciousness is pivotal to theological engagement with the theory of artificial consciousness. We will turn to this subject in the next chapter.

The second aspect of Bavinck's holistic account of the *imago Dei* stresses a collective approach to understanding the meaning of being human. He contests that the meaning of the *imago Dei* is too rich to be fathomed by a single human person; instead, it ties to the whole human community.

> Still, even this creation *in God's image* of man and woman in conjunction is not the end but *the beginning of God's journey with mankind*. It is not good that the man should be alone (Gen. 2:18); nor is it good that the man and woman should be alone. Upon the two of them God immediately pronounced *the blessing of multiplication* (Gen. 1:28). Not the single human person, nor the man and the woman together, but only *the whole humanity is the fully developed image of God*, his children, his race. The image of God is much too rich than that can be fully realized in one single human being, however richly gifted this human being is.[87]

It is clear that, for Bavinck, a collective understanding of the *imago Dei* is essential to the meaning of being human. Although every single human person is the ectype of God, all human beings together can more fully reflect the divine archetype. Furthermore, Bavinck's collective construal strikes a chord with the holistic definition of the *imago Dei* derived from the analysis of Fergusson and Deane-Drummond's revisionist interpretations: the image of God signals the beginning of God's journey with humankind, that is, the commencement of God–human stories.

This collective understanding implies that relationality matters to the *imago Dei*, showing the intimate relationship between human beings.[88] Bavinck describes this human–human relationship as organic. The whole humanity is, so to speak, an organism, which fully realises the *imago Dei*.[89] As such, every single human person is a member of this organism, and all members of this organism are interconnected as a whole. Bavinck's holistic account of the *imago Dei* stops at the human–human relationship. He does not expand upon whether and how the human being as the *imago Dei* is connected with nonhuman others (e.g., animals).

Be that as it may, Bavinck's view of the vestiges of God (*vestigia Dei*) can yield principles for the articulation of a collective understanding of the *imago Dei* in relation to nonhuman others.[90] In concluding his section on the *imago Dei* as the essence of humans, Bavinck writes:

> All creatures are the embodiment of divine thoughts, and all of them exhibit the vestiges of God [*vestigial Dei*]. But all these vestiges, scattered side by side in the spiritual and the material world, are recapitulated in the human being and so interconnected and highly performed that they clearly form the image and likeness of God. The entire world lifts itself upwards, brings itself to an end, completes itself, receives its unity, its goal, and its crown in the human being. In order to be the image of God, the human being, therefore, has to be a recapitulation of the whole nature.[91]

A noteworthy point can be made that, for Bavinck, the *imago Dei* is the foremost and highest vestige of God in creation. At the same time, the *imago Dei* and all other vestiges of God are interconnected and together make the testimony to God in nature. Bavinck's point of view can be further developed to formulate a collective understanding of the *imago Dei*, which involves the relationship between humans and nonhuman others.[92] That is, all vestiges of God are not merely recapitulated in the *imago Dei* but also reinforce its meaning. In this light, Fergusson's and Deane-Drummond's readings of the *imago Dei* complement and enhance Bavinck's holistic account: the *imago Dei* carries the meaning of human performances towards others, which means that the *imago Dei* identifies the whole human person as the storyteller of human ethical performances towards others. Needless to say, this collective understanding of the *imago Dei* paves the way to deal with the question on whether AI is an authentic companion of humans and can be incorporated into the human community. (I will address this question in Chapter 5.)

To sum up, like Fergusson and Deane-Drummond, Bavinck offers a holistic understanding of the *imago Dei*, which highlights both the religious—God–human stories—and the ethical implications—stories of humans and others. Yet, unlike the two contemporary theologians, Bavinck's holistic account rests in the traditional Reformed idea of archetype–ectype: the whole human person as the *imago Dei* is the ectype of God who is the archetype. From this stance are derived the ontological and collective implications of the *imago Dei*.

In subsequent text, it shall be illustrated that holding together these three theologians' holistic stances on the *imago Dei* opens up a way to advance dialogues between theology and AI through crafting a theo-ontology for AI. Specifically, Bavinck's stance on the *imago Dei*—which takes in the theological ontology of archetype–ectype—offers a top-down lens through which the ontological implication of the holistic understanding of the *imago Dei* is brought into the spotlight. In distinction to Bavinck, Fergusson and Deane-Drummond are attentive to evolution and humankind's relationship with other creatures and represent a bottom-up approach to articulating a holistic understanding of the *imago Dei*. These top-down and bottom-up methods converge insofar as Bavinck's connection between religious and moral consciousness and the *imago Dei* corresponds to Fergusson's narrative account of and Deane-Drummond's performative understanding of the *imago Dei*. Their interpretations of the *imago Dei* can be synthesised as follows: *As the signifier of the beginning both of God–human stories and of stories of human ethical performances towards others, the* imago Dei*—which involves both religious and moral consciousness—means that the whole human person as the ectype of the divine archetype consciously performs these religious and moral stories.*

1.4 AI as the Ectype of the Human Being

What role does the archetypal–ectypal interpretation of the *imago Dei* play in a dialogue between theology and AI? As laid out earlier, many scholars generally use one of the three conventional interpretations of the *imago Dei* to illustrate AI and its relationship with humans, which is short of an ontic connection between AI and humans. That is to say, an ontology of AI remains to be developed. On the other hand, the relationship between AI and humans should involve human being as a whole. Hence, the holistic understanding of the *imago Dei* has the potential to bring forth a theologically ontological foundation for human endeavours in AI research. Derived from the archetypal–ectypal model, this ontological foundation means that with respect to the relationship between AI and humans, the human being is the archetype and AI is the ectype.

This archetypal–ectypal understanding of AI–human relationships ontologically ensures the value of AI, which is ultimately grounded in God's creation. Here two questions arise: how can the ontological nomenclature used to describe God-human relationships be employed to define AI–human relationships? Does this methodology imply an identification of humans and AI?

As being created in the *imago Dei*, the human being imitates God's creative activities and, by doing so, exhibits the attributes of the divine archetype. Being created in the *imago Dei* means that humans 'have the prerogative to imitate God … [and] operate as collaborators with God-given resources.'[93] As Michael Burdett remarks, 'Creation is not just God's handiwork alone; we help contribute to it.'[94] From this vantage point, it can be argued that AI is the consequence of the human imitation of God's creation, using God-given resources. Philip Hefner's conception of created co-creator can help us to flesh out human imitative creation. He submits that '[h]uman beings are God's created co-creators whose purpose is to be the agency, acting in freedom, to birth the future that is most wholesome for the nature that has birthed us.'[95] The term 'created co-creator' simultaneously displays the conditionedness (implied in 'created') and freedom (implied in 'co-creator') of human beings.[96] Because human beings are created, they are situated within an ecosystem and a cultural environment.[97] As God's co-creator, human beings 'unavoidably face the necessity both of making choices and of constructing the stories that contextualize and hence justify those choices.'[98] In other words, humans as God's co-creators have the freedom to construct human futures based on the condition of human existence. It stands to the reason that, as one of the most advanced technologies, AI reflects the human effort to construct their futures. As such, the human creation of AI is part of the human identity as God's created co-creator.

Noreen Herzfeld deploys the 'created co-creator' metaphor into her theological engagement with AI by drawing a nuanced distinction between co-*creator* and *co*-creator. She contends that co-*creator* echoes the functional interpretation of the *imago Dei*. AI that is designed in the *imago hominis* is an agent that performs certain tasks in many realms for human beings. *Co*-creator highlights the understanding of the *imago Dei* in terms of relationship. As such, AI that bears the *imago hominis* is viewed as a human companion.[99]

Herzfeld offers an insightful account of how the term 'co-creator' benefits AI-and-theology dialogues. Yet, the 'created co-creator' metaphor cannot be broken up. The meaning of '*created* co-creator' still remains to be uncovered. How does 'created' act as a qualification of 'co-creator' so as to bring to light a deeper meaning of the relationship between humans and AI? The archetype–ectype model fits with and strengthens the 'created co-creator' metaphor while using this metaphor for interdisciplinary studies on AI and theology. That is, AI as the ectype of the human being, who is God's created co-creator, should mirror the *imago Dei*. This ectypal feature of AI conveys three implications.

First, AI is created in the *imago hominis* by the human being, who is the ectype of God and is the created co-creator. As God's co-creator, humans have the freedom to construct their futures on the ground of the condition of their existence. In this light, AI is the fruit of humankind's continuous participation into God's creation. As the ectype of human beings, AI has ontic ties to human beings. On the other hand, given that the idea of archetype–ectype connotes the ontological distinction between God and humans, there must be essential differences between the creative activities of God and humans. While God created all things out of *nothing*, humans create things out of *something*. Whilst speaking of humans as the creator of AI, it is implied that there is the ontological distinction between AI (ectype) and the human being (archetype) due to the fact that AI is not the product of *creatio ex nihilo*.[100] Hence, AI is *not* the *imago Dei* but rather is created in the *imago hominis*.

John Puddefoot's distinction between simulation and emulation can help us to make a conceptual clarification on the ontological difference between AI as the *imago hominis* and human beings as the *imago Dei*. In elaborating on the difference between AI and humankind, Puddefoot alerts us to the different connotations that simulation and emulation carries: 'Simulation … attempts to mimic; emulation attempts to reproduce.'[101] He criticises the indiscriminate use of simulation and emulation in much AI literature, which results in the confusion about mimicking and reproduction. Contrary to such a conceptual confusion, he contends:

> A computer which sets out to do no more than equal or better human performance at some task [sic] could be set up in a way which attempted either to emulate or simulate the performance of the brain. An emulation would have hardware requirements: the computer would need to be built in a way which reproduced brain structure, not just brain function.[102]

As per Puddefoot, emulation presupposes the wholeness of the human being. Both human physical and psychic dimensions need be taken into consideration while speaking of AI as the ectype of the human being in the sense of emulation. However, due to AI's silicon-based system, we can only state that AI is the ectype of carbon-based human beings in the sense of simulation. Many scientists and scholars have paid close heed to the distinction between AI and human beings in view of the latter's biological condition. For example, Margaret Boden stresses that metabolism is a key difference between AI and humans. We will reiterate this difference in chapters that follow, showing how the distinction between the silicon-based system and the biological body lurks in the different implications of AI-powered and human bodily actions in religious activities and moral behaviours. Here it is worth citing the following lengthy passage on the impossibility of reproducing metabolism in AI systems, which serves as a strong case for AI as the ectype in the sense of simulation.

> [Metabolism] can be *modeled* by computers, but not *instantiated* by them. Neither self-assembling robots nor virtual (on-screen) A-Life [artificial life] can actually metabolize. Metabolism is the use of biochemical

> substances and energy exchanges to assemble, and maintain, the organism. So it's irreducibly physical. Defenders of strong A-Life point out that computers use energy, and that some robots have *individual* energy stores, needing regular replenishment. But that's a far cry from the flexible use of interlocking biochemical cycles to build the organism's bodily fabric. So if metabolism is necessary for life, then strong A-Life is impossible. And if life is necessary for mind, then strong AI is impossible too. No matter how impressive the performance of some future AGI, it wouldn't have intelligence, *really*.[103]

Accordingly, an AI system only simulates or mimics, not emulates, the human metabolism. A holistic understanding of the *imago Dei*—which attends both to the psychical and the physical dimensions of the human being—must consider the importance of human biological conditions like metabolism. As such, AI and human beings are ontologically different, albeit that AI can perform many functions in a better way and faster than humans. As will be seen in Chapter 2, this ontological difference between human beings and AI becomes more apparent in the distinction between human and artificial consciousness.

The second implication conveyed by the ectypal feature of AI is the intimate relationship between AI and humans, which means that AI is eligible to participate in the performance of God–human stories. The holistic understanding of the *imago Dei* stresses that the whole human person immerses into God–human stories. For all that AI as the ectype of humanity is designed in the sense of simulation, it occupies a critical role in God–human stories. Our cultural concerns have shown how and to what extent AI may take on a role in God–human stories. For instance, in assessing the religious feature of science fiction, Robert Geraci observes that '[l]iterature reflects the popular concerns of our culture. That intelligent machines present themselves to us like the Holy in science fiction tells us something about how we conceive of them in contemporary life, and this, in turn, tells us something about the nature of modern religious practice.'[104] AI research seems to magnify the religiousness of human beings, and humans seek to reformulate their religion or the idea of God in the age of technology. In a nutshell, the religious elements in our culture, which are partly swayed by AI, uncover the possibility of AI's role in God–human stories.

That said, AI's participation into the God–human story does not mean that AI can occupy a place that is independent of human beings in this story. For AI is not *conscious* of the God–human story in the same way as human beings, which, as will be seen, is determined by the distinction between artificial and human consciousness. When AI is considered as belonging to the God–human story, the ontological difference yet connection between AI and humanity should be foregrounded. I will examine AI's participation in the God–human story in Chapter 3 and discuss questions related to AI-powered liturgical enactment.

Understanding AI as the ectype of the *imago Dei* will give rise to a more positive position towards receiving AI into the faith community, where God–human stories unfold. This can help us to get around some misleading claims made by churches. For example, the report published by the Doctrine Commission of the Church of England in 1987 theologically plays down AI. It raises the concern about the expectation that 'the model of artificial intelligence explains the physical and mental phenomena so fully that it can be regarded as having scientific warrant' for the nature of human personhood.[105] The report is partly correct in that the definition of human nature is not at the disposal of AI research. However, it is overexaggerated to say that all AI research aims to fathom the meaning of human nature.

The Church of Scotland's AI report (May 2021) shows that Christian communities increasingly realise the positive role that AI can play in the faith community.[106] From the perspective of the *imago Dei*, the report illustrates the fact that AI benefits the human community, but, at the same time, it points out ethical issues surrounding the applications of AI (e.g., autonomous cars and killer robots) and the fundamental distinction between AI and humans (e.g., consciousness).[107] 'For the Church,' this report states, 'the place of AI within a bigger narrative of redemption and reconciliation for Creation and relationships, and of its role in the building of the Kingdom of God, should be of concern not just to theologians, but to ministers and congregations.'[108] Hence, it argues that the church should engage with AI through a relational lens such that faith communities can 'move forward with AI as part of their lives.'[109] To this end, this report suggests that the Church needs to explore the opportunities to deploy AI technologies in 'pastoral work, counselling, worship, and mission.'[110]

The Church of Scotland's AI report shows an attempt made to hold together both the functional and relational interpretations of the *imago Dei* to construct an approach to deploying AI in churches. I suggest that the holistic understanding of the *imago Dei* and the idea of AI as the ectype of human beings, as developed in this study, will undergird the Church of Scotland's vision of AI's role in building the Kingdom of God and faith communities, which is bound up with AI's role in God–human stories. (In Chapter 3, I will examine the possible deployment of AI in the Christian liturgy, which echoes this report's suggestion on AI's role in worship.)

The third implication conveyed by the ectypal feature of AI is: AI as the ectype of the human being should reflect the moral nature of the *imago Dei* and mimic the *imago Dei* in a moral way—that is, AI's participation in human ethical performances towards others. As noted earlier, the holistic understanding of the *imago Dei* emphasises the eccentric meaning of being human, that is, being grounded outside ourselves. As an ectype of humans, AI is also embedded with the eccentricity of being human such that AI can simulate the *imago Dei* in its ethical performances towards others. To this extent, AI possesses a moral status. Nevertheless, AI's moral status is derivative and *partial*, which rests with its archetype, the human being. (We will come back to this point in Chapter 4 in exploring the question about AI's moral agency.)

These three implications derived from the ectypal feature of AI demonstrate that AI, as made in the *imago hominis*, is ontologically connected to yet different from human beings who are the *imago Dei*. For this reason, it is misleading to say that AI possesses or is presumed to bear the *imago Dei*. The relationship between the *imago hominis* and the *imago Dei* underlines the value of AI research on the one hand and sets the scene for our exploration of AI from religious and moral perspectives on the other hand.

1.5 Conclusion

The idea of the *imago Dei* can be considered a hinge on which the interaction between theology and AI turns back and forth. This chapter has developed a holistic account of the *imago Dei* to shed light on the ontological relationship between human beings and AI. By doing so, an attempt is made to go beyond engaging theologically with AI research from the perspectives of the three conventional interpretations of the *imago Dei*.

Recent revisionist readings of the *imago Dei* bring to light the importance of the holistic understanding of the *imago Dei*. Fergusson and Deane-Drummond together point us to the contribution that evolutionary theory makes to the holistic understanding of the *imago Dei*. Evolutionary theory helps us to understand that the *imago Dei* is not merely ontologically relevant to the human being but also displays the eccentric value of being human.

The traditional Reformed model of archetype–ectype is employed to expose the ontological foundation of the holistic understanding of the *imago Dei*. In particular, with the archetype–ectype thinking, Bavinck works out a holistic definition of the *imago Dei*, which has a psychosomatic union of the body and the soul as its ontological foundation.

While Fergusson and Deane-Drummond construct a bottom-up approach to the interpretation of the *imago Dei*, Bavinck's hermeneutics is top-down. A holistic definition of the *imago Dei* can be made based on these three thinkers' arguments: the *imago Dei* is the signifier of the beginning of both God–human stories and stories of human ethical performances towards others. This holistic understanding of the *imago Dei* is conducive to offering a theo-ontological foundation for human–AI relationship: as the human being is the ectype of God who is the *uncreated* archetype, AI is the ectype of the human being who is the *created* archetype.

This holistic understanding of the *imago Dei* places emphasis both on the ontological distinction and connection between humans and AI. Following this, it can be inferred that since the holistic understanding of the *imago Dei* carries both religious and ethical connotations, AI as the ectype of human beings is expected to participate into God–human stories and stories of human ethical performances towards others. Given that the holistic interpretation of the *imago Dei* ties religion and ethics to human consciousness, a question arises here: does it mean that AI as the ectype of humanity emulate the *imago Dei* in such a way to perform religious and ethical stories as *consciously* as humans? We are moving on to address this question now.

Notes

1 Coeckelbergh, *AI Ethics* (Cambridge: MIT Press, 2020), 37.
2 David F. Noble, *The Religion of Technology: The Divinity of Man and the Spirit of Invention* (New York: Alfred A. Knopf, 1998), 9–10.
3 Noreen L. Herzfeld, *In Our Image: Artificial Intelligence and the Human Spirit* (Minneapolis, MN: Fortress, 2002), 5.
4 On the history of the interpretation of the *imago Dei*, see J. Wentzel Van Huyssteen, *Alone in the World? Human Uniqueness in Science and Theology* (Grand Rapids: Eerdmans, 2006), 111–162.
5 The substantive understanding of the *imago Dei* has a long history. Theologians, including the likes of Augustine and Thomas Aquinas, are convinced that the *imago Dei* is associated with human rationality, reason, and intellect. Herzfeld's account here is reliant upon Reinhold Niebuhr, *The Nature and Destiny of Man*, 2 vols. (Louisville: Westminster John Knox, 1996).
6 Herzfeld, *In Our Image*, 35–36.
7 Ibid., 40, 73. Herzfeld's emphasis on the importance of human embodiment can also be seen in Noreen Herzfeld, "Bodies Matter: A New Fad and a Fallacy in the Name of Science," in *God's Action in Nature's World: Essays in Honour of Robert John Russell*, ed. Ted Peters and Nathan Hallanger (Aldershot: Ashgate, 2006), 225–233.
8 Nancey Murphy, *Bodies and Souls, or Spirited Bodies?* (Cambridge: Cambridge University Press, 2006); also see Marc Cortez, *Embodied Souls, Ensouled Bodies: An Exercise in Christological Anthropology and Its Significance for the Mind/Body Debate* (London: T&T Clark, 2008).
9 Herzfeld, *In Our Image*, 20–25.
10 Ibid., 41–43.
11 Ibid., 43, 44.
12 Ibid., 44, 45. also see Noreen Herzfeld, "A New Member of the Family? The Continuum of Being, Artificial Intelligence, and the Image of God," *Theology and Science* 5, no. 3 (2007): 240–241; Noreen Herzfeld, "Creating in Our Own Image: Artificial Intelligence and the Image of God," *Zygon: Journal of Religion and Science* 37, no. 2 (2002): 308.
13 Herzfeld, *In Our Image*, 77, 78, 81.
14 Alistair McFadyen, "Redeeming the Image," *International Journal for the Study of the Christian Church* 16, no. 2 (2016): 120.
15 Ibid., 121.
16 Karen O'Donnell, "Performing the *Imago Dei*: Human Enhancement, Artificial Intelligence and Optative Image-Bearing," *International Journal for the Study of the Christian Church* 18, no. 1 (2018): 7.
17 Ibid., 10, 11.
18 Ibid., 9.
19 Ibid., 11.
20 Herzfeld's most recent work comprehensively examines this relational approach to AI. Noreen Herzfeld, *The Artifice of Intelligence: Divine and Human Relationship in a Robotic Age* (Minneapolis: Fortress, 2023).

21 Noreen Herzfeld draws heavily from Karl Barth on the relational reading of the *imago Dei*; Herzfeld, *In Our Image*, 25–31. I will elucidate in Chapter 5 that Barth's understanding of the *imago Dei* cannot be simply considered relational. The recent prominent work on the relational reading is Stanley J. Grenz, *The Social God and the Relational Self: A Trinitarian Theology of the Imago Dei* (Louisville: Westminster John Knox Press, 2001).
22 Noreen Herzfeld, "In Whose Image? Artificial Intelligence and the Imago Dei," in *The Blackwell Companion to Science and Christianity*, ed. J. B. Stump and Alan G. Padgett (Chichester: Wiley-Blackwell, 2012), 506; Herzfeld, *In Our Image*, 45–46; Herzfeld, "Creating in Our Own Image," 310.
23 Herzfeld, *In Our Image*, 47–49.
24 Ibid., 90.
25 Ibid., 92, 93.
26 Herzfeld, "In Whose Image?," 507.
27 Herzfeld, *In Our Image*, 93.
28 Chammah Judex Kaunda, "Bemba Mystico-Relationality and the Possibility of Artificial General Intelligence (AGI) Participation in *Imago Dei*," *Zygon: Journal of Religion and Science* 55, no. 2 (2020): 331.
29 Ibid., 327.
30 Ibid., 334.
31 Ibid., 335.
32 Ibid., 332.
33 Ibid., 333.
34 Ibid., 336, 337.
35 Ibid., 338, 339.
36 Van Huyssteen, *Alone in the World?*, 116.
37 David Fergusson, "Are We Alone? And Does It Matter?: The Narrative of Human Particularity," in *Human Origins and the Image of God: Essaysin Honor of J. Wentzel van Huyssteen*, ed. Christopher Lilley and Daniel J. Pedersen (Grand Rapids: Eerdmans, 2017), 245.
38 David Fergusson, "Humans Created according to the *Imago Dei*: An Alternative Proposal," *Zygon: Journal of Religion and Science* 48, no. 2 (2013): 440.
39 Ibid.
40 Ibid., 441–443; also see David Fergusson, *Creation* (Grand Rapids: Eerdmans, 2014), 11–13.
41 Fergusson, "Humans Created according to the *Imago Dei*," 443–444.
42 Ibid., 444.
43 Fergusson, *Creation*, 14. Fergusson argues that '[w]e have a responsibility to ensure that our way of life does not diminish the natural creation or prevent other creatures from sharing the planet with us.'
44 Fergusson, "Humans Created according to the *Imago Dei*," 445.
45 Ibid., 446. Fergusson develops his argument based on Claus Westermann's exegesis of relevant Genesis passages; see Claus Westermann, *Genesis 1-11: A Continental Commentary* (Minneapolis: Fortress, 1994), 142–161.
46 Fergusson, "Humans Created according to the *Imago Dei*," 446; Celia E. Deane-Drummond, "God's Image and Likeness in Humans and Other Animals: Performative Soul-Making and Graced Nature," *Zygon: Journal of Religion and Science* 47, no. 4 (2012): 934–946.
47 Celia E. Deane-Drummond, *The Wisdom of the Liminal: Evolution and Other Animals in Human Becoming* (Grand Rapids: Eerdmans, 2014), 10–11.
48 Celia E. Deane-Drummond, "In God's Image and Likeness: From Reason to Revelation in Humans and Other Animals," in *Questioning the Human: Toward a Theological Anthropology for the Twenty-First Century*, ed. Lieven Boeve, Yves de Maeseneer, and Ellen Van Stichel (Oxford: Oxford University Press, 2014), 62.

49 Deane-Drummond, *The Wisdom of the Liminal*, 11. She argues in a similar way elsewhere: that humans bear the image of God means that 'they act out divine-image-bearing in a community with God and other creatures.' Deane-Drummond, "In God's Image and Likeness," 75.

50 A helpful comparative study on Fergusson's and Deane-Drummond's understandings of the *imago Dei*, see Eva van Urk-Coster, "Created in the Image of God: Both Human and Non-Human Animals?," *Theology and Science* 19, no. 4 (2021): 343–362.

51 Deane-Drummond, "In God's Image and Likeness," 74.

52 Hence, Deane Drummond argues: 'Human image-bearing becomes in this view not just a bare ontological or relational term but also implies building on such capacities a religious goal to be achieved, an active performance receptive to the grace of God, and a means for creating a community with other creaturely kinds, instead of a means for separation and assumption of superiority.' Ibid., 74.

53 Deane-Drummond, "God's Image and Likeness in Humans and Other Animals," 944; Celia E. Deane-Drummond, *Theological Ethics through a Multispecies Lens*, The Evolution of Wisdom, Volume I (Oxford: Oxford University Press, 2019), 239.

54 Deane-Drummond, "God's Image and Likeness in Humans and Other Animals," 945. It is worth noting that she intends not to blur the distinction between humans and other animals. She argues:

> The difference between other animals and humans is that only the latter are capable of an awareness of their active role in their performance and in this sense are capable of using their minds to adapt to new possible futures. Inasmuch as other animals are not yet perfect, they are still capable of being transformed into a more perfected divine likeness, but this is through the work of grace in eschatological hope rather than in an expectation for present reality.
>
> Deane-Drummond, *The Wisdom of the Liminal*, 307.

55 Deane-Drummond, "In God's Image and Likeness," 72. R. J. Berry suggests that the *imago Dei* can be construed as the shift from *Homo sapiens* to *Homo divinus*. That is, God gave the *imago Dei* to *Homo sapiens* at some time in history so that they evolved as *Homo divinus*. R. J. Berry, "Biology since Darwin," in *Darwinism and Natural Theology*, ed. Andrew Robinson (Newcastle: Cambridge Scholars, 2012), 30. Deane-Drummond takes issue with Berry. She says that 'the idea of God intervening at some point in the history of Homo sapiens seems to go against the sufficiency of God's presence in the world through divine Providence, and to speak of an intervention at some gap-point in our scientific knowledge.' Celia E. Deane-Drummond, "Response: *Homo Divinus*—Myth or Reality?," in *Darwinism and Natural Theology: Evolving Perspectives*, ed. Andrew Robinson (Cambridge: Cambridge Scholars, 2012), 70.

56 Agustín Fuentes, *Why We Believe: Evolution and the Human Way of Being* (New Haven: Yale University Press, 2019), 23.

57 Deane-Drummond, *The Wisdom of the Liminal*, 235.

58 David H. Kelsey, *Eccentric Existence: A Theological Anthropology*, 2 vols. (Louisville, Ky.: Westminster John Knox, 2009), 2:1008; emphasis added.

59 Willem J. Van Asselt, "The Fundamental Meaning of Theology: Archetypal and Ectypal Theology in Seventeenth-Century Reformed Thought," *Westminster Theological Journal* 64, no. 2 (2002): 320–321. On the background to the idea of archetypal and ectypal theology, see Richard A. Muller, *Post-Reformation Reformed Dogmatics, Volume One: Prolegomena to Theology*, 2nd ed. (Grand Rapids: Baker, 2003), 225–228.

60 On the life of Franciscus Junius, see his autobiography: Franciscus Junius, "The Life of the Noble and Learned Franciscus Junius, Doctor of Sacred Theology and

Distinguished Professor at Leiden University," in *A Treatise on True Theology: With the Life of Franciscus Junius* (Grand Rapids: Reformation Heritage Books, 2014), 1–72. On a brief biological sketch of Franciscus Junius and an introduction to his major work, see Willem J. Van Asselt, "Introduction," in *A Treatise on True Theology: With the Life of Franciscus Junius* (Grand Rapids: Reformation Heritage Books, 2014), xi–xlvii.

61 Franciscus Junius, *A Treatise on True Theology: With the Life of Franciscus Junius*, trans. David C. Noe (Grand Rapids: Reformation Heritage Books, 2014), 107.

62 Ibid., 108.

63 Ibid.

64 Ibid., 109.

65 Ibid., 110.

66 Ibid.

67 Ibid., 111.

68 Ibid., 113.

69 Ibid., 117.

70 The Lutheran theologian John Gerhard uses the idea of archetypal and ectypal theology to articulate what true theology is; Johann Gerhard, *On the Nature of Theology and on Scripture*, trans. Richard J. Dinda, ed. Benjamin T. G. Mayes (Saint Louis: Concordia Publishing House, 2009), 22–24. For a helpful analysis of Gerhard's idea of archetypal and ectypal theology, see Robert D. Preus, *A Study of Theological Prolegomena*, The Theology of Post-Reformation Lutheranism, Volume I (St. Louis, Missouri: Concordia Publishing House, 1970), 112–114.

71 For a brief sketch of the development of the idea of archetype–ectype, see Willem J. Van Asselt, "Scholasticism in the Time of Early Orthodoxy (ca. 1560–1620)," in *Introduction to Reformed Scholasticism*, trans. Albert Gootjes (Grand Rapids: Reformation Heritage Books, 2011), 121–126.

72 Dolf te Velde, ed., *Synopsis of a Purer Theology, Volume 1* (Leiden: Brill, 2015), 33.

73 On the life of Francis Turretin, see James T. Dennison Jr, "The Life and Career of Francis Turretin," in Francis Turretin, *Institutes of Elenctic Theology, Volume 3*, trans. George Musgrave Giger, ed. James T. Dennison (Phillipsburg: P&R Publishing, 1992–1997), 639–648.

74 Francis Turretin, *Institutes of Elenctic Theology*, trans. George Musgrave Giger, 3 vols., ed. James T. Dennison Jr (Phillipsburg: P&R Publishing, 1992–1997), 1.2.6.

75 Ibid., 5.10.3.

76 The contemporary Reformed theologian Oliver Crisp uses the notion of archetype–ectype in construing the *imago Dei* Christologically. Oliver Crisp, "A Christological Model of the *Imago Dei*," in *The Ashgate Research Companion to Theological Anthropology*, ed. Joshua Ryan Farris and Charles Taliaferro (Farnham: Ashgate, 2015), 217–229. He argues that 'the divine image you and I bear is, as it were, a facsimile of that image borne by Christ; it is ectypal. Although he lived long after any first putative human community, he is the one who bears the archetypal divine image, after whose divine image the rest of humanity is fashioned' (224).

77 The notion of archetype–ectype is part of the metaphysical foundation for Bavinck's theological anthropology; Brian G. Mattson, *Restored to Our Destiny: Eschatology & the Image of God in Herman Bavinck's Reformed Dogmatics* (Leiden: Brill, 2012), 19–64.

78 Herman Bavinck, *Reformed Dogmatics, Volume 2: God and Creation*, trans. John Vriend, ed. John Bolt (Grand Rapids: Baker, 2004), 530; (hereafter RD2); Herman Bavinck, *Gereformeerde Dogmatiek, Tweede Deel*, 4th ed. (Kampen: J. H. Kok, 1928), 491; (hereafter GD2). The English translation will be revised in places.

79 RD2, 554; GD2, 516.
80 RD2, 532; GD2, 493; emphasis added.
81 RD2, 533; GD2, 493.
82 RD2, 533; GD2, 493.
83 RD2, 555–561; GD2, 517–523.
84 For example, Herman Bavinck, *The Philosophy of Revelation: The Stone Lectures for 1908-1909* (New York: Longmans, Green, and Co., 1909), 77; Herman Bavinck, *Reformed Dogmatics, Volume 1: Prolegomena*, trans. John Vriend, ed. John Bolt (Grand Rapids: Baker, 2003), 520. Cory Brock explains in detail that Bavinck's idea of religion and consciousness takes shape in his critical appropriation of Schleiermacher's conceptual framework of consciousness and feeling while standing in his own Reformed tradition; Cory Brock, *Orthodox yet Modern: Herman Bavinck's Use of Friedrich Schleiermacher* (Bellingham: Lexham, 2020).
85 For example, RD2, 74; Herman Bavinck, *Reformed Ethics, Volume I: Created, Fallen, and Converted Humanity*, ed. John Bolt et al. (Grand Rapids: Baker Academic, 2019), 201, 206, 228.
86 RD2, 70; GD2, 42.
87 RD2, 577; GD2, 538; emphasis added. Bavinck's viewpoint here can be developed to endorse the diversity and unity of the human race; that is, his view of the imago Dei can serve as a foundation for anthropological holism. Jessica Joustra, "An Embodied Imago Dei: How Herman Bavinck's Understanding of the Image of God Can Help Inform Conversations on Race," *Journal of Reformed Theology* 11 (2017): 9–23.
88 Bavinck highlights the relationality as embedded in the imago Dei while explicating the creation of Adam and Eve in the *imago Dei*; Herman Bavinck, *The Christian Family*, trans. Nelson Kloosterman (Grand Rapids: Christian's Library Press, 2012), 3.
89 RD2, 577; GD2, 539.
90 A helpful analysis of Bavinck's view of *vestigia Dei*, see James Eglinton, *Trinity and Organism: Towards a New Reading of Herman Bavinck's Organic Motif* (London: T&T Clark, 2012), 82–84.
91 RD2, 561; GD2, 523–524.
92 The main reason why Bavinck does not develop the idea of the imago Dei in this way is probably his objection to Darwinism and neo-Darwinism; Herman Bavinck, "Evolution," in *Essays on Religion, Science, and Society*, ed. John Bolt, trans. Harry Boonstra and Gerrit Sheeres (Grand Rapids: Baker, 2008), 105–118. Nonetheless, the collective understanding of the imago Dei, being raised above, does not contradict Bavinck's stance insofar as the contemporary evolutionary theory, such as niche construction theory, is immune to his criticism of evolution.
93 John Hedley Brooke, "Visions of Perfectibility," *Journal of Evolution and Technology* 14, no. 2 (2005): 7.
94 Michael Burdett, *Eschatology and the Technological Future* (New York: Routledge, 2015), 242.
95 Philip Hefner, *The Human Factor: Evolution, Culture, and Religion* (Minneapolis: Fortress, 1993), 27.
96 Ibid., 30, 31.
97 Ibid., 36.
98 Ibid., 38.
99 Noreen Herzfeld, "Co-*creator* or *Co*-creator?: The Problem with Artificial Intelligence," in *Creative Creatures: Values and Ethical Issues in Theology, Science and Technology*, ed. Ulf Görman, Willem Drees, and Hubert Meisinger (London: T&T Clark, 2005), 45–52.
100 This observation turns down Gregory Peterson's critique of Hefner's use of the prefix co- in 'co-creator.' Peterson argues that this prefix 'implies not simply

that we are creating in and of our own right but that our creative acts are in cooperation with God's creative acts in a way that suggests partnership rather than subordination.' Gregory R. Peterson, "The Created Co-Creator: What It Is and Is Not," *Zygon: Journal of Religion and Science* 39, no. 4 (2004): 829. In conjunction with the archetype–ectype model, this metaphor emphasises that human creative acts are in subordination to God's creative acts.

101 John C. Puddefoot, *God and the Mind Machine: Computers, Artificial Intelligence, and the Human Soul* (London: SPCK, 1996), 66.

102 Ibid.

103 Margaret A. Boden, *AI: Its Nature and Future* (Oxford: Oxford University Press, 2016), 144–145.

104 Geraci, "Robots and the Sacred in Science and Science Fiction," 976.

105 The Doctrine Commission of the Church of England, *We Believe in God* (London: Church House, 1987), 47. Yet many Christian scholars dispute with this report's stance on AI, e.g., Fraser Watts and Mark Williams, "The Doctrine Commission and Artificial Intelligence," *Theology* 91, no. 741 (1988): 216–217.

106 The Church of Scotland, *Artificial Intelligence (AI): Opportunity and Challenge for the Church* (May 2021), https://www.churchofscotland.org.uk/__data/assets/pdf_file/0011/79760/Artificial-Intelligence-SRT-report-22.2.21WEB.pdf. Access by 20 October 2021.

107 Ibid., 5.1–6.2, 8.1–8.7.

108 Ibid., 10.1.

109 Ibid., 10.5–6.

110 Ibid., 4.7.

2 The Consciousness of the *Imago Hominis*

> I do not think we can exclude the possibility of conscious computers on metaphysical grounds, but there may be empirical grounds for the impossibility of computer consciousness.
>
> ———Ian Barbour[1]

> [The Machine] fulfils the dream of all living creatures by satisfying our instinctive craving for the maximum of consciousness.
>
> ———Teilhard de Chardin[2]

Five decades ago, Hubert Dreyfus argued in the first edition of *What Computers Still Can't Do* (1972) that 'the term "intelligence" can be misleading. No one expects the resulting robot to reproduce everything that counts as intelligent behavior in human beings.'[3] Dreyfus's pessimistic position towards artificial intelligence (AI) invited much criticism from AI researchers at the time.[4]

Dreyfus's assertion would confront with more challenges in the contemporary age because of the rapid progress in AI research, particularly the development of artificial neural networks (ANNs), which is also named connectionism. ANNs model the networks of neural cells in the human brain and are composed of a large number of interconnected nodes (processing units), each of which only computes one thing.[5] On 20 January 2022, it was reported that AI researchers at Meta AI have successfully worked out Data2vec, 'a single algorithm that can be used to train a neural network to recognize images, text, or speech.'[6] It is hard to deny that the simultaneous performance of these three skills is, in some sense, an intelligence behaviour. The advocators of artificial consciousness raise more challenges to Dreyfus's stance. In particular, the proponents of artificial general intelligence (AGI) and artificial superintelligence (ASI) are convinced that human-level AI will be created in the future and that it will *consciously* act in the same way as humans.

DOI: 10.4324/9781003356738-3

AI's consciousness (artificial consciousness or machine consciousness) has become one of the most contentious areas in the field of philosophy and ethics of AI. In relevant AI literature, many scholars have examined the idea of artificial consciousness, and the (im)possibility of artificial consciousness is intensively being debated from both technical and humanistic perspectives. (This chapter will look into some of these studies.) However, few of them seek to explore the ontological underpinning of consciousness. In the previous chapter, I have demonstrated that the theo-ontological foundation, which is constructed with the holistic account of the *imago Dei* in conjunction with the Reformed theology of archetype–ectype, should be laid out so as to distinguish between the human being as archetype and AI as ectype, between the *imago Dei* and the *imago hominis*. In this chapter, I will unpack the distinction between the *imago Dei* and the *imago hominis* with an eye to bringing to light artificial consciousness as the consciousness of the *imago hominis*.

I will argue that being made in the *imago hominis*, AI as the ectype of human beings has a *simulated* consciousness. Specifically, the holistic understanding of the *imago Dei* entails the corollary that the idea of consciousness is bound up with human biological conditions to such an extent that an *emulated* consciousness cannot be conceived without emulating the human being as a biological creature, especially the neural system in the human brain.

I proceed to explore the variegated accounts of consciousness, critically examining the notions of phenomenal and access consciousness, which have been widely employed to account for AI's consciousness. Following this, I will analyse the idea of creature consciousness and point out that the idea of the *imago Dei* theo-ontologically grounds creature consciousness. Through this analysis, it can be seen that despite the essential distinction between human and artificial consciousness, AI is conducive to the human *conscious* narration of both God–human stories and stories of human ethical performances towards others.

2.1 Variegated Accounts of Artificial Consciousness

It does not hold true for all scholars that the question about consciousness is critical to AI research. 'Consciousness,' Drew McDermott argues, 'is only marginally relevant to artificial intelligence … because to most researchers in the field other problems seem more pressing.'[7] Some AI researchers' indifference to consciousness is partly due to the ambiguous definition of this term. René Descartes (1596–1650) is championed by some as the first philosopher who developed a systematic approach to the study of consciousness, illustrating the correlation of consciousness, the human brain, and nonconscious processes.[8] However, up until now, there is no consensus on the definition of 'consciousness,' but various forms of consciousness are delineated to show the characteristic features of human consciousness.

Riccardo Manzotti and Vincenzo Tagliasco contest that the precise definition of consciousness is not necessary in technology because terminological ambiguity disappears when a physical artefact is created and operates in the real world.[9] In other words, the function of AI-powered artefacts will reveal the core meaning of artificial consciousness. Yet, they are oblivious to the fact that it is weak AI that works properly in the real world and yet triggers debates over artificial consciousness. Even some proponents of AGI and ASI envisage AI's human-level consciousness or mind machine through the lens of weak AI insofar as they insist that weak AI will continue to evolve until the actualisation of human-level consciousness. Hence, it is necessary to clarify the term 'consciousness' while exploring artificial consciousness.

Before defining the term 'consciousness,' John Puddefoot's caveat on the Sliding Definition Ploy is worth noting. He contends that, in exploring the idea of mind machine, some scholars 'agree to use a word in a precise but rather unusual sense while we are on the alert, but subtly shift the meaning later to embrace everyday senses of the word not covered by their initial definition.'[10] This reminds us of the importance to investigate whether the meanings of 'consciousness' in human and artificial consciousness are kept consistent throughout studies on humans and AI. In this chapter, I will use a univocal 'consciousness' while speaking of human and artificial consciousness and, by doing so, I intend to underline some philosophical presuppositions that give rise to the Sliding Definition Ploy in relevant literature on artificial consciousness.

Although consensus has not yet been reached regarding the definition of consciousness, Michael Tye's taxonomy of consciousness can help us to capture some nuanced understandings of consciousness in relation to artificial consciousness and mind machine. According to Tye, consciousness can be classified into three categories. Phenomenal consciousness refers to mental states of subjective experiences, including perceptual experiences, bodily sensations, felt reactions, and felt moods. Creature consciousness means identifying the human *being* as being conscious of things. Access consciousness connotes some mental states that are related to certain kinds of information processing.[11] These three types of consciousness display their distinct philosophical orientations and emphases. Phenomenal consciousness is empirical, laying emphasis on what is experienced and felt, whereas access consciousness focuses on epistemology, making consciousness bound up with information alone. By contrast, creature consciousness concentrates on ontology, that is, on the *being* of the conscious agent. Variegated accounts of artificial consciousness fall into one of these three categories of consciousness. I will first examine phenomenal and access consciousness, which, by and large, are characteristic of most views of artificial consciousness.

2.1.1 *Phenomenal Consciousness*

The idea of phenomenal consciousness is very common in AI literature. Manzotti and Tagliasco contest that phenomenal consciousness is foundational to artificial consciousness.[12] Early in 1988, based on the idea of phenomenal consciousness, Hans Moravec registered that AI's development will eventually and inevitably produce artificial consciousness. He contended that the programmes developed in his lab have been able to answer the questions like 'Where are you?' and 'Why did you do that?'. The answers made by the programmes show senses of space and direction, which means that the programmes can *perceive* surrounding circumstances. As such, he claimed that the mind machine's consciousness at the time will evolve into a human-level consciousness.[13] For all that Moravec did not use the term 'phenomenal consciousness,' it is explicit that he endorsed the notion that the programmes in his lab were able to perceive and obtain subjective experiences of the real world in the same way as human beings.

The idea of subjective experience is not uncommon in the discussion over phenomenal consciousness, and these two notions are even considered identical in some works.[14] The idea of subjective experience lies at the base of the Hard Problem of Consciousness. David Chalmers notes: 'If any problem qualifies as the problem of consciousness, it is this one. In this central sense of "consciousness," an organism is conscious if there is something it's like to be that organism, and a mental state is conscious if there is something it's like to be in that state.'[15] The Hard Problem of Consciousness stems from a presupposition that subjective experience is distinct from the processes of the human brain. However, the latest philosophical, theological, and scientific studies on consciousness turn down such a presupposition. While speaking of subjective experience, what is presupposed is the (human or artificial) agent's interaction with the external world and this interaction is tied up with the human brain. (We will turn to examine this point of view in the subsection 2.2.) This point of view is appreciated by many proponents of artificial consciousness.

Riccardo Manzotti and Sabina Jeschke recently explored the causal interaction between the human being and the external world, and they sought to justify the idea of artificial consciousness. They take issue with the identification between consciousness and the outcome of the processing of human internal cognitive structures.[16] Rather, they argue that consciousness lies at the foundation of cognitive structures and is the consequence of the causal correlation between an agent's body and the external world. 'Consciousness is not something produced by an agent (no matter whether inside in terms of computational processes or outside in terms of actions), but rather it is the kind of basic causal structure between the body of the agent and the environment.'[17] In this light, consciousness must be phenomenal insofar as it rests in what is experienced by the agent's body in the real world. For Manzotti and Jeschke, this understanding of consciousness opens up a way to bring into existence a *conscious* artificial agent. That is, genetic algorithms—e.g., programmes that can change and improve themselves by adapting to changing environments—can be used to construct a conscious artificial architecture, which produces variants of the artificial agent's brain structure.[18] With the redefined notion of phenomenal consciousness, Manzotti and Jeschke indiscriminately apply the notion of consciousness to both artificial and human agents. Some AI scholars disagree with the notion of artificial consciousness like the one raised by Manzotti and Jeschke because it blurs the boundaries between the artificial and the natural.[19] Be that as it may, Manzotti and Jeschke rightly attend to the interaction between the physical body and the real environments in the construal of consciousness. We will turn to this subject anon.

Another issue raised by Manzotti and Jeschke's conscious artificial architecture is: despite justifying the idea of qualia, they endorse the claim that the artificial agent can have qualia in that the causal interaction between the body and the environment can genuinely happen in AI.[20] Qualia refer to qualities and properties of experiences felt by the human being. Hence, qualia are subjective and ineffable, and exist inside the agent. Some scholars are reluctant to consider qualia in AI research because it is subjective and philosophically defined. Ron Chrisley, for example, argues that the issue of qualia 'should be decided empirically/experientially; it is not the place of philosophy to prejudge the issue.'[21] Marcus Hunter also contends that consciousness and qualia may be 'forever beyond formalization.'[22]

However, echoing Manzotti and Jeschke's emphasis on the role that the human body plays in the function of consciousness, Daniel Dennett and Marcel Kinsbourne argue, 'Conscious experiences are real events occurring in the real time and space of the brain, and hence they are clockable and locatable within the appropriate limits of precision for real phenomena of their type.'[23] Following this, the qualities and properties of conscious experiences can be justified by virtue of their temporality and locality. Qualia as the properties of conscious experiences are inextricably connected to something that really exists. This reality of qualia is reinforced by David Longinotti's recent study on qualia and biological life. He notes that 'the source of qualia is some sort of living substance.' Qualia are produced through 'an energy transduction' that takes place in particular neurons, and such transduction can be verified by some measurement of the energy consumed in the human brain.[24] Longinotti's claim reflects his attention to the close tie between the human brain and consciousness as well as subjective experience and, in turn, undergirds the real connection between the subjective conscious agent and external circumstances. That being so, scholars will do injustice to the idea of qualia if they reject it only because qualia are subjective. On the contrary, as John Searle contests, the 'problem of consciousness is identical with the problem of qualia, because conscious states are qualitative states right down to the ground. Take away the qualia and there is nothing there. ... Conscious states by definition are inner, qualitative, subjective states of awareness or sentience.'[25]

A question arises: is it possible, as Manzotti and Jeschke suggest, that AI or an artificial agent has qualia? John Puddefoot's question is worth noting:

> what does it mean to say that the computer tastes the scrambled egg? This is the *qualia* problem. It is a source of great embarrassment to the AI community, and their response is either to deny that the problem exists or to argue it away. These are both minimization strategies. That a computer can be designed with circuits capable of identifying a substance as scrambled egg is not in dispute. That identifying that substance as scrambled egg is *the same thing* as tasting it, is very much in dispute, and is in fact the crux of the whole AI debate.[26]

Puddefoot's observation lays bare a critical issue in articulating artificial consciousness based on the idea of phenomenal consciousness. That is, the idea of phenomenal consciousness is often caught up into the morass of empiricism such that consciousness is enslaved to the external world. To put it another way, if consciousness is merely phenomenal, then it is always the external world that imposes representations on the agent's consciousness. One can then raise the questions: Does consciousness have a subjective dimension? Is the prefix 'self' in self-consciousness still justified? Viewed as such, the notion of phenomenal artificial consciousness ironically contradicts its own accentuation on subjective experience. The notion of access consciousness shows several attempts that have been made to address these two questions.

2.1.2 Access Consciousness

With an emphasis on information processing, many proponents of artificial consciousness—understood in the sense of access consciousness—often adopt computational models of consciousness. Aaron Sloman and Ron Chrisley argue that consciousness refers to information processing in humans and animals precisely because, in order to survive, all organisms have to act through information selection under certain conditions and require sorts of information processing systems, which fit with themselves according to their own conditions. Moreover, organisms process information based on their architecture, which includes, *inter alia*, forms of representation, algorithms, concurrently active sub-systems, connections between these sub-systems. This architecture-based consciousness can explain *qualia*.[27]

Sloman and Chrisley's study shows a commonality of access consciousness: the consciousness of both AI and human beings is, in essence, a computer, and operates by computing data. With this in mind, Benjamin Kuipers boldly argues: 'The essential features of consciousness can, in principle, be implemented on a robot with sufficient computational power and a sufficiently rich sensorimotor system, embodied and embedded in its environment.'[28] To be sure, Kuipers is convinced that artificial consciousness is nothing other than a computer and that its actualisation relies upon computational power and calculation speed. This is reminiscent of Turing's claim that a machine that can think will be created when a digital computer has 'an adequate storage capacity as well as working sufficiently fast.'[29]

ANNs are the most common computational model that can be used to construe artificial consciousness. Traditional AI is designed with symbolic programmes, and symbolic AI depends on the sequential operations of a set of symbols and instructions. By contrast, ANNs aim to get rid of the mechanical feature of symbolic AI by *modelling* the system of biological neurons in the human brain.[30] Attempts are made to construct an artificial system that consists of a large number of interconnected nodes which can process information in a dynamic and parallel way. 'Sequential instructions,' Margaret Boden notes, 'are replaced by massive parallelism, top-down control by bottom-up processing, and logic by probability. And the dynamical, continuously changing aspect of ANNs contrasts starkly with symbolic programs.'[31]

ANNs seem to open up a way to *reproduce* biological neural systems and thus underpin some scholars' optimistic position towards the possibility of artificial consciousness. Of these scholars, Ray Kurzweil is a prominent proponent of AGI and artificial consciousness. He contends:

> We will have a massive neural net, built from silicon and based on a reverse engineering of the human brain. Its circuits are a million times faster than human neurons, and it can learn human language and model human knowledge. It develops its own conceptions of reality, and on its own blurts out that it is lonely. We will ask whether this robot is a conscious agent. In the end we will come to believe that such robots are conscious much as we believe the same of each other. Their minds will be based on the design of human thinking, and they will embody human qualities and claim to be human. And we'll believe them.[32]

Kurzweil is convinced that artificial consciousness will be actualised by replicating the neural system of the human brain. This conviction rests in the claim that consciousness depends upon neurons alone. The approach to using ANNs to create artificial consciousness presupposes this understanding of the connection between consciousness and neurons. Although it is an exaggeration to say that neurons are the only condition for consciousness, an element of truth nevertheless lies in this stance: consciousness is not utterly passive in such a way that it only receives the representations imposed by the external world; rather, it involves subjective factors of an agent. But what are these subjective factors?

Antonio Chella, Marcello Frixione, and Salvatore Gaglio propose a computational model which can be used to explain these subjective factors. By doing so, they seem to raise an approach to artificial self-consciousness. By 'self-consciousness,' they refer to AI robot's reflection on 'itself, its own perceptions and

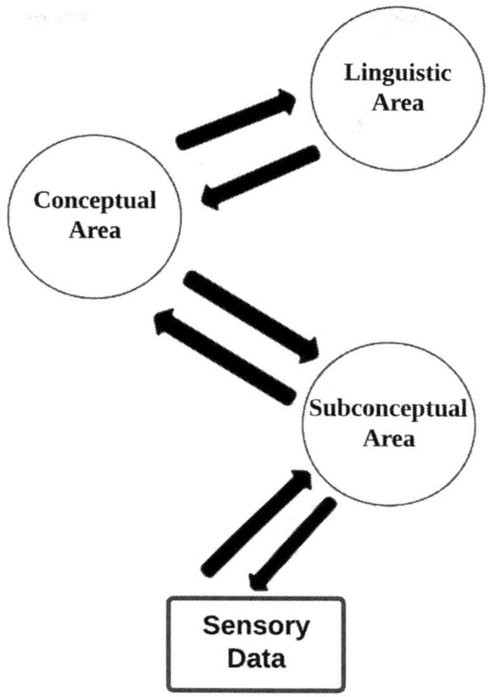

Figure 2.1 The cognitive architecture for AI's self-consciousness[34]

actions during its operating life.'[33] The three scholars have designed a cognitive architecture, as Figure 2.1 shows, which is composed of three areas.

The subconceptual area stores basic perceptual data that are obtained through sensors. The data in the subconceptual area are categorised in the conceptual area, which consists of conceptual spaces. A conceptual space, which generates a number of knoxels (points in a conceptual space), serves as a connection point between subconceptual and linguistic areas. Based on a logic-oriented formalism, the processing and representations of the categorised data arise in the linguistic area. According to Chella, Frixione, and Gaglio, what happens in the linguistic area will generate the self of a robot. That is, basic perceptual data are transformed into logic representations, which is retrospective in the sense that the AI robot can retrieve older knoxels.[35] In this sense, they identify knoxels as subjective factors of artificial self-consciousness.

Access consciousness is typical of Chella, Frixione, and Gaglio's cognitive architecture. Based on data processing, the three scholars aim to strike a balance between the subjective factors of the agent and the objective phenomenal world. Needless to say, Chella, Frixione, and Gaglio's cognitive architecture serves as a case against one-sided emphasis either on the phenomenal world or on the (biological or artificial) agent itself.

The idea of access consciousness, to a certain extent, offers more comprehensive construal of consciousness with computational models than the idea of phenomenal consciousness. The advocates of access consciousness attend to the function and structure of the human brain while articulating artificial consciousness, which in turn brings to explicit the biological condition of consciousness. Morphological computing is a case in point, reflecting the concern about the human biological condition in the construction of computational models. Gordana Dodig-Crnkovic observes that morphological computing—which stresses the constraint of a (human or artificial) agent's morphology on its interactions with the environment and on its own development—is dependent upon morphogenesis, which refers to 'the biological process that causes an organism to develop its shape.'[36] In her mind, this morphogenesis is tied up with informational computation precisely because the 'fabric of reality' is nothing other than information. 'Reality consists of informational structures organized on different levels of abstraction/resolution.'[37] This being the case, both biology and (AI) robotics are, in essence, involved with morphological computing, that is, an agent's informational computation and self-structuring according to its own morphology. As per Dodig-Crnkovic, morphological computing is not an analogy or metaphor but is a *model* that has 'substantial non-linguistic, interactive spatio-temporal and visual qualities.'[38] Hence, she suggests that morphological computing is conducive to the development of cognition and intelligence in (AI) robots.[39]

Morphological computing falls into the category of computational models. Although Dodig-Crnkovic does not use morphological computing to endorse artificial consciousness, the univocal idea of morphogenesis in biology and robotics underwrites that a human-level artificial consciousness is likely to be created through informational computation.[40]

What is at issue in the idea of access consciousness is nothing other than computational models. It can be seen in ANNs, Chella, Frixione, and Gaglio's cognitive architecture, and morphological computing, that constructing artificial consciousness in the sense of access consciousness is predicated upon modelling the structure and function of the human brain. Of course, it should be conceded that the human consciousness or mind does have a sort of computing power and procedure, which can be clearly seen in human mathematical calculations and logical reasoning. Nevertheless, it is a simplistic approach to fully realising this computing power and procedure through simulating the structure of the human brain, which *de facto* means to utterly identify the human brain with consciousness.

Reductive physicalism or materialism is characteristic of this simplistic approach. Computational models of consciousness materialise consciousness and reduce nonphysical properties of human consciousness to physical particulars and properties, which reflects the continuing influence of eighteenth-century materialism. Ann Thomson's extensive study shows that in the eighteenth century, despite drawing our attention to the fact that thoughts rest upon the material brain, materialism fails to explain how the structure and working of the human brain could produce thoughts.[41] As a result, the nonphysical is reduced to the physical. Over the past several decades, nonreductive physicalism—which opposes the reduction of nonphysical properties to physical particulars—has offered a more convincing argument that irreducible nonphysical consciousness and soul are closely related to human physical particulars.[42] As will be illustrated, this close connection cannot be easily created with computational models.

Furthermore, relying upon the notion of information processing, the idea of access consciousness remains subject to Puddefoot's critique of the Sliding Definition Ploy. As Hubert Dreyfus rightly notes, the meaning of 'information processing' is vague.

> If this term simply means that the mind takes account of meaningful data and transforms them into other meaningful data, this is certainly incontrovertible. But the cybernetic theory of information … has nothing to do with meaning in this ordinary sense. It is a nonsemantic, mathematical theory of the capacity of communication channels to transmit data.'[43]

This terminological ambiguity of 'information processing' largely incurs the indiscriminate application of AI's computational models to the human brain. If AGI and ASI proponents claim that access consciousness is human-level consciousness in that AI processes information in the same way as humans, then they fail to disambiguate the term 'information processing' used respectively in the cases of AI and humans and consequently leave out the limitations of computational models.

2.1.3 *Summary*

Through the above analysis, it can be argued that as mirrored in the above different accounts of artificial consciousness, the ideas of phenomenal consciousness and access consciousness are short of a sound holistic understanding of the human being. The notion of phenomenal consciousness presupposes that the whole human being is subordinate to the phenomenal world. As such, human subjectivity is called into question. The notion of access consciousness gives so much attention to the structure and operation within the human brain that it is susceptible of downplaying the nonphysical and psychic dimensions of the human being. Be that as it may, we cannot draw the conclusion that neither phenomenal consciousness nor access consciousness has anything to do with human consciousness. The point I make here is: if human consciousness is defined *only* in the sense of either phenomenal consciousness or access consciousness or both, it can be argued that human-level artificial consciousness will be actualised. Yet, the questions raised above challenge this corollary. The accounts of these two types of consciousness reveal that an ontological account of consciousness, namely, creature consciousness, is in demand.

2.2 The Ectype of Creature Consciousness in *Imago Hominis*

2.2.1 Creature Consciousness

Unlike the ideas of phenomenal consciousness and access consciousness, the notion of creature consciousness stresses the being of humans and describes the human agent as psychosomatic in articulating consciousness. From this it follows that creature consciousness is tied up with humanity itself and carries heavy ontological weight, which draws our attention to its distinction from both phenomenal and access consciousness. At the same time, this ontological weight can respond to Ned Block's ambiguity theory of consciousness. Block maintains that the term 'consciousness' is 'a mongrel concept' and thus needs to be ambiguous.[44] With an emphasis on the function of consciousness, he attempts to make a precise distinction between phenomenal consciousness and access consciousness. He sums up three main differences. First, the content of phenomenal consciousness is phenomenal, whereas that of access consciousness is representational. Second, contrary to phenomenal consciousness, access consciousness 'is a functional notion' and thus 'is system-relative.' Third, 'there is such a thing as a [phenomenal conscious] type or kind of state,' but access consciousness is informational and its state is temporary.[45] Following these observations, Block asserts that conceptually speaking, access consciousness and phenomenal consciousness can operate without each other.[46]

Block does not spell out how and to what extent phenomenal consciousness and access consciousness are correlated. I suggest that creature consciousness serves as an ontological foundation for both phenomenal and access consciousness, and consequently integrates the latter two into the idea of the human being as the *conscious* agent. Neil Manson's distinction between state consciousness—which is relevant to mental states—and creature consciousness can serve to elucidate the latter's ontological import. As with Michael Tye, he contends that creature consciousness justifies the usage of consciousness in the portrayal of creatures. At the same time, he stresses that state consciousness pertains to the mental states of the agent rather than to the agent itself.[47] Hence, creature consciousness is conceptually broader than state consciousness insofar as a conscious creature can have both conscious and unconscious mental states.[48] From this it follows that both phenomenal consciousness and access consciousness are subsumed under state consciousness, that is, the agent's conscious mental states. As such, the notions of both phenomenal consciousness and access consciousness presuppose the being of the conscious agent. Since neither phenomenal nor access consciousness is embedded with ontological implications, creature consciousness serves as an ontological prerequisite for both phenomenal consciousness and access consciousness.[49]

From the vantage point of a human agent as a conscious being, it comes to be seen that the notion of creature consciousness is characterised as holistic, seeking to articulate the meaning of consciousness based on a psychosomatic account of human beings. While phenomenal consciousness is entangled with the external world and access consciousness is preoccupied with the computational model of the human brain, the notion of creature consciousness occupies itself with the human being *per se*, including both physical and psychical dimensions.

This psychosomatic trait of creature consciousness is in tune with the holistic account of the *imago Dei* as analysed in the Section 1.2 of Chapter 1. The *imago Dei* refers to the signifier of the beginning of both God–human stories and stories of human ethical performances towards others. As such, the *imago Dei* means that the psychosomatic human person, as the ectype of the divine archetype, *consciously* performs these religious and moral stories. These conscious performances of religious and ethical stories, making up the eccentricity of human beings, flow from the very being of humans precisely because God created human beings as such. These conscious performances, by their very nature, rest neither with phenomenal consciousness nor with access consciousness. Rather, the conscious performance connoted by the *imago Dei* is made on an ontological basis that is more radical than state consciousness. In a nutshell, the human agent's conscious performances of these religious and moral stories originate from their being and essence, from the *imago Dei*, and from their identity as the ectype of the divine archetype. Both the *imago Dei* and creature consciousness point us to the psychosomatic union of the human person while elaborating on consciousness.

2.2.2 *Consciousness and the Brain*

The ontological and holistic emphases embedded within the ideas of both creature consciousness and the *imago Dei* sit well with recent studies on the relationship between mind/consciousness and the human brain. For example, Foresight Mental Capital and Wellbeing (2008), a project overseen by a high-level stakeholder group and advised by scientific experts, states in its final report that both 'our mental and material resources' are vital.[50] It points out that the popular understandings of the brain, such as 'left brain' versus 'right brain' in education, which are spread through social media, are misleading. Instead, the relationship between the human brain and mind should be examined for the improvement of the education of children.[51] It is clear that recent studies on the relationship between consciousness and the brain lend support to a holistic anthropology beneath the concept of creature consciousness, which, in turn, sheds lights on how AI's consciousness—if it is loosely understood—can only be construed from the perspective of the *imago hominis*. I proceed to engage with John Searle's philosophy of consciousness and Sarah Lane Ritchie's theology of consciousness before demonstrating that artificial consciousness should be considered to be ectypal.

2.2.2.1 John Searle's Philosophy of Consciousness

John Searle's extensive analysis of the brain and consciousness is conducive to the construction of a philosophical apparatus through which to use the notion of creature consciousness to clarify that artificial consciousness should be construed from the perspective of the *imago hominis* rather than the *imago Dei*. At the same time, his studies can philosophically flesh out the meaning of creature consciousness.

The 'central fact of specifically human existence' is, according to Searle, tied up with consciousness, which underpins human emotions, language, and all other activities.[52] Unlike some scholars who use consciousness and mind interchangeably, he distinguishes between these two terms. By the term 'mind,' he refers to 'the sequences of thoughts, feelings and experiences, whether conscious or unconscious, that go to make up our mental life.'[53] In this sense, consciousness is a part of human mental life and denotes a state of the mind.

Much attention is drawn, in Searle's work, to the role that the human brain plays in the functioning of consciousness. Yet, he does not follow the line of reductive physicalism to reduce nonphysical properties of human consciousness to physical particulars and properties of the brain. While stressing the importance of human biological conditions, Searle describes consciousness as a feature of the brain derived from neural activities: 'consciousness and indeed all mental phenomena are caused by lower-level neurobiological processes in the brain,' and 'consciousness and other mental phenomena are higher-level features of the brain.'[54]

It is apparent that Searle opposes a zero-sum approach to the physical and the mental aspects of human beings. Neither the mental nor the physical aspects of humanity can simply be reduced to each other. Instead, both should be equally considered while investigating human consciousness. On the one hand, the brain and consciousness should not be mixed up: 'The brain processes do not, at the level of neuron firings, constitute consciousness; rather, the neuron firings at the lower level cause the higher-level or system feature of consciousness.'[55] On the other hand, he cautions against the rupture between the brain and consciousness: 'Our conscious states are higher-level or system features of the brain, and consequently there are not two separate sets of causes—the psychological and the neurobiological. The psychological is just the neurobiological described at a higher level.'[56] This unconfused and unseparated relationship between the brain and consciousness displays Searle's understanding of consciousness in terms of holistic anthropology. That is, like the notions of the *imago Dei* and creature consciousness, Searle's philosophy of consciousness underscores that underpinning consciousness is the idea of the human being as a psychosomatic whole.

This anthropological holistic interpretation of consciousness gives shape to the ontological understandings of subjective experiences, feelings, intentionality, cognition, and other mental activities. In terms of Tye's taxonomy of consciousness, Searle's account of consciousness falls into the category of creature consciousness, bringing to light the ontological principles of his elaboration on phenomenal and access consciousness.

Searle suggests that consciousness is ontologically subjective by virtue of the experience of human *subject*.[57] Consciousness is bound up with all aspects of human life and 'names the mode in which humans ... conduct the major activities of their lives.'[58] One crucial feature of consciousness is intentionality. By 'intentionality,' Searle means the 'capacity of the mind by which mental states refer to, or are about, or are of objects and states of affairs in the world other than themselves.'[59] In other worlds, all experience of the human agent, which is associated with her activities in the world, is characterised by intentionality. (We will return to intentionality in relation to liturgical enactment and moral agency in Chapters 3 and 4.) Searle asserts that the connection between the human agent's intentional mental life and bodily contact with the physical world consists in her brain. 'Visual and auditory experiences, tactile sensations, hunger, thirst, and sexual desire, are all caused by brain processes and *they are realised in the structure of the brain*, and they are all intentional phenomena.'[60] In this light, the *biological* nature of the brain is essential to the actualisation of intentionality in the activities of human lives in the physical world, and carbon-based neurobiological process in the brain is a biological foundation of human subjective experience.

Through the above analysis, it comes to be seen that Searle's view of subjective experience is radically divergent from that the notion of artificial consciousness articulated in the sense of phenomenal consciousness. Subjective experience cannot be narrowed down to such an extent that it only denotes something that is passively imposed. On the contrary, by virtue of the biological nature of the human agent's brain, experience is, first of all, subjective.

Searle's objection to the notion of artificial consciousness is intensified by his critique of access consciousness. Again, his emphasis on the biological basis of consciousness plays a significant role here. Through exploration of the idea of information processing, he contends:

> The problem with the concept of 'information processing' is that information processing is typically in the mind of an observer. For example, we treat a computer as a bearer and processor of information but, intrinsically, the computer is simply an electronic circuit. We design, build, and use such circuits because we can interpret the inputs, outputs, and the intermediate processes as information bearing. But in such a case the information in the computer is in the eye of the beholder, it is not intrinsic to the computational system. ... The reason we cannot analyse consciousness in terms of information processing and symbol manipulation is that *consciousness is intrinsic to the biology of nervous systems*; information processing and symbol manipulation are observer-relative.[61]

Searle makes it clear that the essential distinction between human consciousness and artificial consciousness consists in that the silicon-based system of information processing, which is designated as artificial consciousness, is created without intrinsic neurobiological processes. To put it differently, artificial consciousness will become possible when, and only when, the human biology of nervous systems can be fully reproduced in the computational system.

Searle's elucidation on consciousness brings into light how *creature* consciousness serves as the ontological foundation for phenomenal and access consciousness. In view of the human being as a psychosomatic whole, he opens up a way for us to deploy the idea of *imago hominis* into the conceptual articulation of artificial consciousness, which buttresses theological engagement with AI research. More specifically, as has been illustrated, with the model of archetype–ectype and the holistic understanding of the *imago Dei*, AI can be construed as the ectype of humanity and made in the *imago hominis*. According to Searle's analysis of consciousness, the functioning of consciousness in AI systems—which are identified by scholars as either phenomenal consciousness or access consciousness—is not involved with the whole human being and consequently fails to prove that AI is conscious of things. While speaking of AI as made in the *imago hominis*, what is underlined is the particularity of the biological and physical properties of human beings, especially the neurobiological processes in the brain, which can never be reduced to binary systems and algorithms. Following the line of Searle's argument, the idea of the *imago hominis* does not connote that AI as the ectype of humanity emulates the *imago Dei* in such a way to perform religious and ethical stories as *consciously* as humans. Rather, the *imago hominis* indicates that artificial consciousness is the ectype of and ontologically different from human consciousness.

2.2.2.2 *Sarah Lane Ritchie's Theological Analysis of Consciousness*

In her celebrated *Divine Action and the Human Mind*, Sarah Lane Ritchie offers a theological analysis of consciousness while delineating the relationship between divine action and the human mind. Her elaboration on consciousness is grounded on the in-depth engagement with and critique of Philip Clayton—who ontologically severs consciousness from the neural activities of the brain—and, by doing so, seeks to justify the openness of the human mind to divine action.[62]

Ritchie's criticism of Clayton's stance rests in her sentiments both that consciousness does not lies outside the inquiry of the natural sciences and that its *physicality* should not be renounced. Following this she intends to constructively develop a theistically naturalistic theory of consciousness. It is worth noting that Ritchie defines the terms, 'naturalism,' 'nature,' and 'natural' in broad senses while taking issue with both metaphysical and ontological naturalism. The former sets natural science as the only guide and the point of departure for our philosophy of nature, whereas the latter pits the natural against the supernatural.[63] As per Ritchie, a broader sense of naturalism can reserve a space for the supernatural and divine action and get rid of scientism. Her own methodology is designated as 'theistic naturalism', that is, 'the theological affirmation that a full account of nature would necessarily include an account of God's active, immanent involvement in nature.'[64] In this way, the so-called conflict between divine action and consciousness is methodologically erased.

Ritchie seeks to naturalise the human mind. 'By arguing that theistic naturalism (broadly construed) redefines nature in theological terms such that all physicality is inherently involved with God's active presence, [Ritchie] removes the theological need to privilege the mind as the spiritual part or aspect of a person.'[65] Following this, she embarks on exploring three theistically naturalistic approaches to divine action, which are relevant to our understanding of consciousness and can shed light on the difference between human and artificial consciousness.

First, Ritchie examines Thomistic approaches to divine action. Thomistic divine action emphasises a God-world model that describes God and natural agents as working on differing levels. This view of double agency—God as the primary cause and natural agents as secondary causes—underwrites that 'all nature is something of a causal joint; to pinpoint one particular aspect of nature as open to divine action is erroneous.'[66] Thomistic naturalistic approach to divine action echoes Ritchie's assertion that the human mind is not privileged in regard to divine action.[67] Moreover, Ritchie maintains that 'there is literally nothing that the scientific community could say that would challenge Thomism; Thomistic divine action is framed in such a way that it automatically incorporates all scientific knowledge.'[68] Through the Thomistic naturalistic lens it can be seen that the naturalisation of the mind indicates that human consciousness should be theologically examined in the purview of nature. This being the case, human consciousness cannot be conceptualised without our *natural* bodies, that is, biological bodies. Again, the close tie between human consciousness and biological conditions is put in the foregrounded.

Second, Ritchie explores a version of panentheistic naturalism in reference to divine action. This panentheistic naturalism—which is grounded on panentheism that the world exists in God who is more than the world—is developed by Orthodox theologian Christopher C. Knight to counter a sort of deism that makes God far away from nature.[69] It affirms that God's interference with nature makes nature more natural precisely because divine actions *per se* are higher laws of nature.[70] This panentheistic naturalism helps theologians not to identify the mind as nonphysical so as to justify divine action.[71] In this light, the physicality of the mind is defended, and the close connection between consciousness and the human body is retained and made prominent. At the same time, this panentheistic naturalism undergirds the religious character of consciousness. That is, since divine action makes natural consciousness more natural, human consciousness *in se* is ever open to God.

Third, Ritchie investigates pneumatological naturalism and contends that it provides a great theological framework for divine action. 'Pneumatological naturalism,' she asserts, 'is a version of theistic naturalism that sees the Spirit's presence and influence in nature as normative—it is natural for nature to participate in, and experience the influence of, the Spirit.'[72] In terms of pneumatological naturalism, the human mind is not uniquely nonphysical locus of divine action but rather is 'a particularly intense locus of human participation in God.'[73] While panentheistic naturalism brings to light the religious character of human consciousness, pneumatological naturalism underscores the particular role that human consciousness plays in religion.[74] Ritchie is attentive to how pneumatological naturalism crafts a framework in which the human brain operates in a particular manner. The following lengthy passage is worth quoting:

> In pneumatological naturalism … divine actions are conceived as variously intense manifestations of the Spirit who is always present to begin with at all levels within nature. The intensity of these manifestations may be largely influenced by the attention, response, and disposition of creatures participating in and with the Spirit. While this is largely speculative at this point, one could imagine that the brain's immense complexity would enable humans to naturally respond to and cooperate with God in a more intense manner than would, for example, a pine tree. … the remarkable complexity of the human mind-brain may simply allow for a comparatively enhanced ability to participate in the Spirit and be affected by God in a way that may seem miraculous.[75]

Accordingly, pneumatological naturalism gives due attention both to the mind and the body while elucidating divine action in nature and, in this way, is grounded in a holistic anthropology and accentuates the physicality of consciousness.

To sum up, Ritchie's exploration of divine action in the human mind theologically fleshes out the ontological meaning of creature consciousness. Neither reductive physicalism nor physical–spiritual dualism can do justice to the fact that the whole human being is a conscious agent. Needless to say, Ritchie's stance squares with the holistic understanding of the *imago Dei* in that both are rooted in a holistic meaning of being human. Viewed as such, she alerts us to the physicality of consciousness in articulating artificial consciousness. Her study also compels us to pay close heed to the religious character of human consciousness while dwelling on the idea of artificial consciousness. Therefore, whether artificial consciousness is counted as creature consciousness is determined by the extent to which divine action operates in consciousness in the same way as described by theistic naturalism. (This is particularly relevant to Chapter 3, which will explore AI's application in liturgical enactment.)

2.2.3 *Is Artificial Consciousness Religious and Moral?*

Searle and Ritchie together shed light on the priority of the ontological meaning of consciousness with emphasis on the indissoluble bond between consciousness/mind and the human body. Since this ontological implication is psychosomatic, they respectively offer philosophical and theological justification for creature consciousness as the ontological foundation for both phenomenal and access consciousness. Malcolm Jeeves's study on the interaction between the brain and human spirituality is a case in point to bolster such an ontological foundation and show the spiritual particularity of human consciousness. Based on the findings and development of mind–brain studies in the past several decades, he contends:

> Changes in our brains may be reflected in changes in our spiritual awareness. There are well-documented changes in the spiritual dimensions of life, loosely called spirituality, which are clearly dependent upon the intactness and the normal functioning of the neural substrates of behavior. … The spiritual dimension to life is embodied. At the same time we must recognize the importance of cognitive processes such as beliefs and expectations, as well as the social context in which these are held, and which underline the importance of recognizing that spirituality is firmly embedded in contexts of relationships both horizontal with one another and vertical with the God we worship.[76]

Jeeves's observation expressly places the whole human being at the centre of spirituality, showing that spiritual life involves both mental and physical aspects of humanity. What humans are phenomenally and informationally conscious of in spiritual life is ontologically predicated upon the being of humans.

The ontological implication of creature consciousness brings the idea of artificial consciousness full circle to the meaning of AI as the ectype of the human being and as made in the *imago hominis*. The question about artificial consciousness boils down to whether the *imago hominis* has creature consciousness. The holistic understanding of the *imago Dei* and the physical underpinning of creature consciousness together underline the significance of the *carbon-based biological* human body for human consciousness. That being so, we can logically arrive at the corollary that ectypal artificial consciousness that is human-level consciousness should be connected to a *carbon-based biological* body that emulates, not simulates, the human body. Granted, we need to concede that, *metaphysically* speaking, the state-of-the-art silicon-based AI may evolve into a system with the biological body identical to that of the human being. Be that as it may, Ian Barbour notes:

> conscious awareness requires forms of organized complexity or properties of neural cells and networks that have no parallels in silicon-based systems. *I do not think we can exclude the possibility of conscious computers on metaphysical grounds, but there may be empirical grounds for the impossibility of computer consciousness.* Because we know so little about the physical basis of human consciousness or the directions of future research in computer science, I am willing to leave this question open.[77]

The theo-ontological foundation of AI, as laid out in the preceding chapter, aligns with Barbour's observation in this passage. For all that one could be metaphysically open to the possibility of human-level artificial consciousness, such theo-ontological foundation of AI—epitomised by the claim that AI as made in the ectype of the *imago Dei*—casts this metaphysical optimistic attitude towards AGI and ASI into doubt insofar as the latter two do injustice to human biological conditions which are empirically constitutive of being human.

With this being said, artificial consciousness qualifies the ectype of creature consciousness. While there is the ontological distinction between artificial and creature consciousness, AI as made in the *imago hominis* means that some functions of human consciousness are simulated within AI, illustrating the similarities between artificial and human consciousness as revealed through the operation of phenomenal and access consciousness. In this connection, the idea of artificial consciousness as the ectype of human consciousness raises the question about the sense in which artificial consciousness is religious and moral. In the preceding chapter, I have demonstrated that the theological thinking of archetype–ectype ontologically undergirds the claim that the religious and moral natures of human consciousness can be *simulated* (not emulated) in AI. However, a question still lingers: Is there any AI research that echoes this theo-ontological claim?

Recent studies on the motion of artefacts can lend support to the claim that artificial consciousness as the ectype of creature consciousness is made up of the features and qualities that human beings attribute to AI. In analysing Searle's philosophy of consciousness, it has been demonstrated that intentionality is a critical feature of consciousness. That is to say, an AI robot's intentional action showcases that artificial consciousness can be viewed as attributed consciousness. Kazunori Terada, Takashi Shamoto, and Akira Ito's study on human–robot interaction makes it evident that humans attribute their characteristics to the movements of artefacts.[78] The three scholars attend to how humans attribute intention to artefacts under the influence of different contexts. In their experiment, 60 university students are, respectively, invited to observe a chair that moves along B-spline trajectory, which is remotely controlled by an experimenter. There are four experimental conditions: (1) the chair moves along the B-spline trajectory; (2) a desk is in the experimental area while the chair is moving along B-spline trajectory; (3) the chair moves towards the desk every five minutes while moving along B-spline trajectory; (4) there is no desk in the experimental area, but the chair moves to the place where the desk was place in condition (3) every five minutes. Based on the questionnaires completed by these students, it is found that eight, five, five and twelve participants, respectively, in each experimental condition perceive the motion of the chair to be goal-oriented. The experimental result is made that the participants attribute the intentionality to the chair according to its movement profile, regardless of the physical appearance. Therefore, Terada, Takashi, and Ito remark that 'making humans treat an artifact as an intentional existence is important for humans to construe and interpret its actions. However, it is not necessary for artifacts to actually have intention or a mental state.'[79] In similar manner, AI users can attribute certain intentions to AI actions and identify AI as an agent that acts according to its *consciousness* of contexts.

The three researchers' experiment serves to demonstrate how humans would attribute intentionality and consciousness alike to AI (robots). If AI's actions are viewed as religious and moral, then it is humans who attribute their religious and moral intentions to AI. All the more salient is that AI designers and programmers can attribute their own religious and moral intentions—which are closely related to their consciousness—to AI and robots that they create. I proceed to go into a comprehensive consideration of details related to the religious and moral issues surrounding the AI designer's involvement into the creation of artificial consciousness. At this point, I will briefly discuss the attribution of consciousness to AI from a methodological perspective.

A hitherto reiterated metaphysical principle is that the ontological distinction between the *imago Dei* and the *imago hominis* lies at the foundation of artificial consciousness as the ectype of human consciousness. Aside from this ontological endorsement of the ectypal feature of artificial consciousness, Jobst Landgrebe and Barry Smith present us with a methodological analysis of the impossibility of human-level artificial consciousness. In their recent work, Landgrebe and Smith distinguish between three sorts of impossibility: the technical, the physical, and the mathematical.[80]

Technical impossibility means that the purpose cannot be achieved due to the technology we currently have. By implication, what is now technically impossible may be achieved with technical breakthroughs and innovations in the future. Hence, technical impossibility is provisional and expected to be converted to technical possibility. Physical impossibility refers to the impossibility caused by the contravention of the laws of physics. Yet, it has been well demonstrated that the laws of physics are asymptotic approximations for physical reality rather than universally prescriptive of physical reality itself.[81]

In this light, physical impossibility is also provisional due to the scientifically limited description of nature. Mathematical impossibility connotes that 'some mathematically specified problems' cannot be solved due to '*a priori* reasons of mathematics.'[82] As noted earlier, AI is engineered with algorithms (mathematical operations) to model functions performed by the human mind. Landgrebe and Smith's objection to AGI as well as human-level artificial consciousness relies upon the mathematical impossibility concomitant with AI.

> *The model is not exactly isomorphic to the aspect of reality it represents.* Rather it is made through a process of conscious selection from a slice of reality by using abstraction. Most models are not about the individuals which make up the relevant slice of reality, but rather about the corresponding universals.... The abstraction used in creating models means that they represent idealisations of reality.[83]

While technical impossibility may give one some hope for AGI or human-level artificial consciousness in the future, mathematical impossibility rules out such hope insofar as mathematical models underpin potential breakthroughs that may take place in the sphere of AI research.

That said, Landgrebe and Smith's idea of mathematical impossibility sheds methodological light on the meaning of artificial consciousness as the ectype of human consciousness. Inasmuch as models created for AI are the idealisations of reality, artificial consciousness can be considered the consciousness of the *imago homini*, which flows from the simulation of the consciousness of the *imago Dei*. As such, artificial consciousness and human consciousness are ontologically correlated yet distinguished through the lens of the *imago Dei*. In this sense, artificial consciousness can be considered religious and moral in an ectypal sense.

2.3 Conclusion

Following the theo-ontological foundation of AI laid in Chapter 1, I have expanded, in this chapter, on this theo-ontological foundation through exploration of artificial consciousness. That is, artificial consciousness is the ectype of human consciousness and is attributed with religious and moral characteristic features (e.g., intentionality). The ontological distinction between human and artificial consciousness lies in the particularity of creature consciousness. Human biological conditions can never simply be reduced to information processing or phenomenal perception. In this connection, a great deal of scholarly attention has been drawn to the role that the human brain, and neural activities therein, play in the operation of human consciousness. We will return to the connection between neural activities and the human mind time and again in the chapters that follow.

Considering artificial consciousness as the ectype of human consciousness has the potential to advance the ontological examination of the religious and moral issues surrounding AI applications. Since the *imago Dei* is the signifier of the beginning of both God–human stories and stories of human ethical performances towards others, AI as the ectype of the human beings can consciously (in an ectypal sense) perform God–human stories and stories of human ethical relationship with others. In this sense, AI serves to expand and extend the human performances of religious and ethical stories. In the following chapters, I shall explore how AI participates into the performance of human religious and ethical stories.

Before unfolding the religious and moral features of AI, it is worth making a clarification on the precedence of Chapter 3 over Chapters 4 and 5. Chapter 3 will examine the role of AI in the Christian liturgical enactment. Chapters 4 and 5 will focus on AI's moral status and companionship. The order of chapters is determined by the holistic understanding of the *imago Dei* as developed in Chapter 1. Drawing on Bavinck's work, I have demonstrated that the *imago Dei* is correlated with consciousness. While investigating AI ontologically through the doctrinal lens of the *imago Dei*, we need to attend to how the religious and moral features of consciousness *in se* interplay with each other. Bavinck's explication of the relationship between religion and morality is of importance for figuring out the interconnection between religious and moral features of consciousness.

Bavinck argues that the Decalogue comprises two tables: the first table states the relationship between God and humans or human duties towards God (religion), and the second table formulates human moral duties.[84] Hence, he asserts that 'religion is to morality ... what the first table is to the second table.'[85] At the same time, Bavinck contends that religion must come before morality. This is so because '[d]uties toward God take precedence over duties toward ourselves and the neighbor, family, country, government, and so forth ... The commandments of the first table come before those of the second table.'[86]

Following Bavinck's view of the Decalogue's two tables, it comes to be seen that the religious feature of consciousness takes precedence over the moral feature. Artificial consciousness as the ectype of human consciousness is embedded with such a sequence. As such, I shall first explore AI's role in the Christian liturgy, followed by the inquiry into AI's moral status and companionship. Now we turn to address the question concerning AI-powered liturgical enactment vis-à-vis liturgical enactment performed by human clerics.

Notes

1 Ian G. Barbour, "Neuroscience, Artificial Intelligence, and Human Nature: Theological and Philosophical Reflection," *Zygon: Journal of Religion and Science* 34, no. 3 (1999): 379.

2 Pierre Teilhard de Chardin, *The Future of Man*, trans. Norman Denny (New York: Image Books, 2004), 239.

3 Hubert L. Dreyfus, *What Computers Can't Do: A Critique of Artificial Reason* (New York: Harper & Row, 1972), xxvi. This claim remains unchanged in the revised edition; Hubert L. Dreyfus, *What Computers Still Can't Do: A Critique of Artificial Reason*, revised ed. (Cambridge, Massachusetts: MIT Press, 1992), 78.

4 See Pamela McCorduck, *Machines Who Think: A Personal Inquiry into the History and Prospects of Artificial Intelligence* (Natick, Massachusetts: A K Peters, 2004), 211–239.

5 For a brief history of ANN and its difference from symbolic AI, see Ron Sun, "Connectionism and Neural Networks," in *The Cambridge Handbook of Artificial Intelligence*, ed. Keith Frankish and William M. Ramsey (Cambridge: Cambridge University Press, 2014), 108–127.

6 Will Douglas Heaven, "Meta's New Learning Algorithm Can Teach AI to Multi-Task," *MIT Technology Review* 20 January 2022, https://www.technologyreview.com/2022/01/20/1043885/meta-ai-facebook-learning-algorithm-nlp-vision-speech-agi/.

7 Drew McDermott, "Artificial Intelligence and Consciousness," in *The Cambridge Handbook of Consciousness*, ed. Philip David Zelazo, Morris Moscovitch, and Evan Thompson (Cambridge: Cambridge University Press, 2007), 117.

8 On Descartes's view of consciousness and his role in the history of the study of consciousness, see Chris D. Frith and Geraint Rees, "A Brief History of the Scientific Approach to the Study of Consciousness," in *The Blackwell Companion to Consciousness*, ed. Susan Schneider and Max Velmans (Chichester: Wiley Blackwell, 2017), 3–16; also see William Seager, "A Brief History of the Philosophical Problem of Consciousness," in *The Cambridge Handbook of Consciousness*, ed. Philip David Zelazo, Morris Moscovitch, and Evan Thompson (Cambridge: Cambridge University Press, 2007), 9–33.

9 Riccardo Manzotti and Vincenzo Tagliasco, "Artificial Consciousness: A Discipline between Technological and Theoretical Obstacles," *Artificial Intelligence in Medicine* 44 (2008): 106. Other AI scholars, such as Robert Clowes and Anil Seth, argue that contemporary AI research has not yet developed a commonly accepted framework of artificial consciousness. In this sense, AI-powered artefacts do not remove the terminological ambiguity in the idea of artificial consciousness; Robert W. Clowes and Anil K. Seth, "Axioms, Properties and Criteria: Roles for Synthesis in the Science of Consciousness," *Artificial Intelligence in Medicine* 44 (2008): 91–104.

10 Puddefoot, *God and the Mind Machine*, 57.

11 Michael Tye, "Philosophical Problems of Consciousness," in *The Blackwell Companion to Consciousness*, ed. Susan Schneider and Max Velmans (Chichester: Wiley Blackwell, 2017), 17–18. Ron Chrisley summarises the different categorisations of artificial consciousness, which are raised by other scholars; Ron Chrisley, "Philosophical Foundations of Artificial Consciousness," *Artificial Intelligence in Medicine* 44 (2008): 121–125. The merit of Tye's taxonomy lies in its attention to the philosophical dimension of consciousness.

12 Manzotti and Tagliasco, "Artificial Consciousness," 110–111.

13 Hans P. Moravec, *Mind Children: The Future of Robot and Human Intelligence* (Cambridge, Massachusetts: Harvard University Press, 1988), 39.

14 For example, Chuanfei Chin, "Artificial Consciousness: From Impossibility to Multiplicity," in *Philosophy and Theory of Artificial Intelligence 2017*, ed. Vincent C. Müller (Cham: Springer, 2018), 3–18; Stan Franklin, "IDA: A Conscious Artifact?," *Journal of Consciousness Studies* 10, no. 4–5 (2003): 47–66.

15 David Chalmers, "The Hard Problem of Consciousness," in *The Blackwell Companion to Consciousness*, ed. Susan Schneider and Max Velmans (Chichester:

Wiley Blackwell, 2017), 33; also see David Chalmers, "Facing Up to the Problem of Consciousness," in *Explaining Consciousness—The "Hard Problem"*, ed. Jonathan Shear (Cambridge, Massachusetts: MIT Press, 1997), 10.

16 They aim to turn down the computational model of consciousness, which will be spelt out below.

17 Riccardo Manzotti and Sabina Jeschke, "A Causal Foundation for Consciousness in Biological and Artificial Agents," *Cognitive Systems Research* 40 (2016): 174–175.

18 Ibid., 180–182.

19 For example, Stevan Harnad and Peter Scherzer, "First, Scale up to the Robotic Turing Test, Then Worry about Feeling," *Artificial Intelligence in Medicine* 44 (2008): 83–89.

20 Manzotti and Jeschke, "A Causal Foundation for Consciousness in Biological and Artificial Agents," 178.

21 Chrisley, "Philosophical Foundations of Artificial Consciousness," 136.

22 Marcus Hutter, "One Decade of Universal Artificial Intelligence," in *Theoretical Foundations of Artificial General Intelligence*, ed. Pei Wang and Ben Goertzel (Paris: Atlantis, 2012), 71.

23 Daniel C. Dennett and Marcel Kinsbourne, "Time and the Observer: The Where and When of Consciousness in the Brain," *Behavioral and Brain Sciences* 15 (1992): 235.

24 David Longinotti, "Agency, Qualia and Life: Connecting Mind and Body Biologically," in *Philosophy and Theory of Artificial Intelligence 2017*, ed. Vincent C. Müller (Cham: Springer, 2018), 49–50. He makes an attempt to describe the production of qualia as follows: 'Living matter contains potential energy that resides in its structure, an organization of atoms and molecules that enables the substance to perform the activities that keep it alive. I hypothesize that action potentials can damage the organizational structure of the q-source. As neural activity begins to have this effect, the q-source reacts by converting the electro-chemical pulses to feelings. Ridding itself of the neural spikes in this way enables the q-source to return to its original structure, which benefit is compensated by the immediate expenditure of the qualia. Energy is thereby conserved; the energy in the neural pulses is converted to the energy of the feeling, which is instantly exchanged for the potential energy of the q-source. Accordingly, the generation of feelings is an action of the q-source that is due to its homeostatic nature; qualia are produced by living matter of some kind' (53–54).

25 John R. Searle, "How to Study Consciousness Scientifically," in *Consciousness and Human Identity*, ed. John Cornwell (Oxford: Oxford University Press, 1998), 28.

26 Puddefoot, *God and the Mind Machine*, 28; emphasis added.

27 Aaron Sloman and Ron Chrisley, "Virtual Machines and Consciousness," *Journal of Consciousness Studies* 10, no. 4–5 (2003): 133–172.

28 Benjamin Kuipers, "Drinking from the Firehose of Experience," *Artificial Intelligence in Medicine* 44 (2008): 155.

29 Turing, "Computing Machinery and Intelligence," 38–39.

30 For example, Wen-Chuang Chou and Tsung-Ying Sun use a computational model to simulate the human neural system and construct a perception pattern corresponding to that in the brain. Wen-Chuang Chou and Tsung-Ying Sun, "The Computational Model to Simulate the Progress of Perceiving Patterns in Neuron Population," in *Artificial Neural Networks: Biological Inspirations – ICANN 2005*, ed. Włodzisław Duch et al. (Berlin: Springer, 2005), 7–12.

31 Boden, *AI*, 79.

32 Raymond Kurzweil, *The Age of Spiritual Machines: When Computers Exceed Human Intelligence* (New York: Viking, 1999), 9; also see Ray Kurzweil, *The*

Singularity is Near: When Humans Transcend Biology (London: Viking Penguin, 2005), 378–379.

33 Antonio Chella, Marcello Frixione, and Salvatore Gaglio, "A Cognitive Architecture for Robot Self-Consciousness," *Artificial Intelligence in Medicine* 44 (2008): 148.

34 Ibid.

35 Ibid., 151–153.

36 Gordana Dodig-Crnkovic, "The Info-computational Nature of Morphological Computing," in *Philosophy and Theory of Artificial Intelligence*, ed. Vincent C. Müller (Berlin: Springer, 2013), 60. She describes this biological process as follows: 'During the process of its development, based on its DNA code, the body of a living organism is created through morphogenesis, which governs the formation of life over a short timescale, from a single cell to a multi-cellular organism, through cell division and organization of cells into tissues, tissues into organs, organs into organ systems, and organ systems into the whole organism.'

37 Ibid., 62.

38 Ibid., 66.

39 Ibid.

40 Nonetheless, I agree with Dodig-Crnkovic that morphological computing can 'advance our understanding of biological life.' Ibid., 66.

41 Ann Thomson, "Materialistic Theories of Mind and Brain," in *Between Leibniz, Newton, and Kant: Materialistic Theories of Mind and Brain*, ed. Wolfgang Lefèvre (Dordrecht: Springer, 2001), 149–173. It is worth noting that Thomson analyses both atheistic materialism and the materialism that is sympathy with Christianity. Yet, both of them attempt to explain natural phenomena only from the perspective of matter.

42 E.g., Nancey Murphy, "Nonreductive Physicalism: Philosophical Issues," in *Whatever Happened to the Soul?: Scientific and Theological Portraits of Human Nature*, ed. Warren S. Brown, Nancey Murphy, and H. Newton Malony (Minneapolis: Fortress, 1998), 127–148; Nancey Murphy, "Physicalism Without Reductionism: Toward a Scientifically, Philosophically, and Theologically Sound Portrait of Human Nature," *Zygon: Journal of Religion and Science* 34, no. 4 (1999): 551–571; Warren S. Brown, "Nonreductive Physicalism and Soul: Finding Resonance Between Theology and Neuroscience," *American Behavioral Scientist* 45, no. 12 (2002): 1812–1821. However, nonreductive physicalist accounts of consciousness is not unproblematic *in se*, not least because the contradiction between reductionism and physicalism is not thoroughly settled. On a critical engagement with nonreductive physicalist ideas of consciousness, see Sarah Lane Ritchie, *Divine Action and the Human Mind* (Cambridge: Cambridge University Press, 2019), 175–183.

43 Dreyfus, *What Computers Still Can't Do*, 165.

44 Ned Block, "On a Confusion about a Function of Consciousness," *Behavioral and Brain Sciences* 18 (1995): 236.

45 Ibid., 232.

46 Ibid., 233–235.

47 Neil Manson, "State Consciousness and Creature Consciousness: A Real Distinction," *Philosophical Psychology* 13, no. 3 (2000): 405–406.

48 Ibid., 407.

49 Gualtiero Piccinini maintains that creature consciousness 'is constitutive of phenomenal consciousness'; Gualtiero Piccinini, "The Ontology of Creature Consciousness: A Challenge for Philosophy," *Behavioral and Brain Sciences* 30, no. 1 (2007): 103–104.

50 Foresight Mental Capital and Wellbeing, *Mental Capital and Wellbeing: Making the Most of Ourselves in the 21st Century*, The Government Office for Science

(London, 2008), 9. https://assets.publishing.service.gov.uk/government/uploads/system/uploads/attachment_data/file/292450/mental-capital-wellbeing-report.pdf

51 Ibid., 127–130.

52 John R. Searle, *Minds, Brains and Science* (Cambridge, Massachusetts: Harvard University Press, 1984), 16.

53 Ibid., 10–11.

54 Searle, "How to Study Consciousness Scientifically," 21; Searle, *Minds, Brains and Science*, 19.

55 John R. Searle, *Mind: A Brief Introduction* (Oxford: Oxford University Press, 2004), 149. He reformulates this argument as follows: 'consciousness is caused by microlevel processes in the brain and realized in the brain as higher-level or system feature' (158).

56 Ibid., 227.

57 Searle, "How to Study Consciousness Scientifically," 26. Searle also argues, 'Conscious states by definition are inner, qualitative, subjective states of awareness or sentience.' (28)

58 Ibid., 32.

59 Searle, *Mind*, 28.

60 Searle, *Minds, Brains and Science*, 24; emphasis added.

61 Searle, "How to Study Consciousness Scientifically," 36; emphasis added.

62 Ritchie, *Divine Action and the Human Mind*, 81–130.

63 Ibid., 201–207.

64 Ibid., 232.

65 Ibid., 191.

66 Ibid., 235–240.

67 It should be noted that Ritchie critically engages with Thomistic naturalism, pointing out that the questions about double agency and the causal joint problem are not fully solved; ibid., 246–252.

68 Ibid., 260.

69 Ibid., 263; Christopher C. Knight, *The God of Nature: Incarnation and Contemporary Science* (Minneapolis, MN: Fortress, 2007), 116.

70 Ritchie, *Divine Action and the Human Mind*, 282–287.

71 Ibid., 289.

72 Ibid., 299–300.

73 Ibid., 300.

74 As such, whilst elaborating on pneumatological naturalism, Ritchie seeks to 'draw together the threads of various theistic naturalisms, suggesting that while they face significant critiques, theistic naturalism may offer a theologically robust and scientifically compatible approach to divine action in the mind, relatively speaking.' Ibid.

75 Ibid., 330–331.

76 Malcolm Jeeves, "Brains, Minds, Souls, and People: A Scientific Perspective on Complex Human Personhood," in *The Depth of the Human Person: A Multidisciplinary Approach*, ed. Michael Welker (Grand Rapids: Eerdmans, 2014), 98.

77 Barbour, "Neuroscience, Artificial Intelligence, and Human Nature," 379; emphasis added.

78 Kazunori Terada, Takashi Shamoto, and Akira Ito, "Human Goal Attribution toward Behavior of Artifacts," in *17th IEEE International Symposium on Robot and Human Interactive Communication* (IEEE, 2008), 160–165.

79 Ibid., 164. A recent study on AI's attributed intention, see Davide Ghiglino and Agnieszka Wykowska, "When Robots (pretend to) Think," in *Artificial Intelligence: Reflections in Philosophy, Theology, and the Social Sciences*, ed. Benedikt Paul Göcke and Astrid Rosenthal-von der Pütten (Paderborn: Mentis, 2020), 53–62.

80 Jobst Landgrebe and Barry Smith, *Why Machines Will Never Rule the World: Artificial Intelligence Without Fear* (New York: Routledge, 2023), 8–9.

81 Further on this point, see Robert E. Ulanowicz, "The Universal Laws of Physics: Inflated Ontologies?," in *God's Providence and Randomness in Nature: Scientific and Theological Perspectives*, ed. Robert John Russell and Joshua M. Moritz (West Conshohocken: Templeton Press, 2018), 69–84.

82 Landgrebe and Smith, *Why Machines Will Never Rule the World*, 8–9.

83 Ibid., 109–110; emphasis added. They intend not to play down the value of models used for AI. 'On the contrary, their validity—and thus their usefulness—rests precisely on their ability to abstract. This abstraction reveals aspects of reality subject to universally valid laws' (110).

84 Bavinck, *Reformed Ethics, Vol. I*, 56, 201, 222.

85 Ibid., 62.

86 Herman Bavinck, *Reformed Ethics, Volume II: The Duties of the Christian Life*, ed. John Bolt et al. (Grand Rapids, Michigan: Baker Academic, 2021), 84.

3 The Worship of the Church and the AI-Powered Enactment of the Liturgy

The holistic construal of the *imago Dei* is made up of both religious performances towards God (God–human stories) and ethical performances towards others (stories of humans and others). Being made by a human agent who is in psychosomatic unity, this double performance underlies the meaning of the whole human person as created in the *imago Dei*. The down-to-earth aspects of this double performance are multifaceted. Geoffrey Wainwright's scriptural and theological inquiry into Christian worship helps us to focus on two particular aspects of this performance: the *imago Dei* means both communion with God in liturgy and social relations with others.[1] In Chapter 5, I will discuss social relations and artificial intelligence (AI) through exploration of the sense in which AI is a companion. In this chapter, the subject matter is AI-powered liturgy examined in terms of the ontological distinction yet connection between the *imago Dei* and the *imago hominis*, echoing the main claim made in Chapter 1 that the *imago Dei* identifies human beings as participants in the performance of God–human stories.

The term 'liturgy' comes from the Greek word λειτουργία, which is composed of λεῖτος (public) and ἔργον (work). The term λειτουργία originally refers to public service (e.g., civil service) related to particular communities. In terms of religion, liturgy refers to the public worship of faith communities, be it script-based, largely spontaneous, or in other styles. 'It may very well draw upon,' Frank Senn notes, 'symbols that are regarded as natural in that they are shared by many groups of people in different times and places …, but the community of faith also gives expression to its own worldview and values by use of these symbols.'[2] As such, one religion can have variegated liturgies, multiple rituals (formal and regularly performed ceremonies), and liturgical texts, which largely depend upon historical contexts and faith communities. Be that as it may, for Christianity, there are theological principles that are grounded in the doctrine of the incarnation and are common to liturgies across churches of different theological traditions. For the purpose of this study, I will concentrate on these common principles mindful of liturgies varying across different theological traditions. Such liturgical diversity-in-unity will shed light on the limitations of AI-powered liturgy and how AI may have a bearing on liturgies.

DOI: 10.4324/9781003356738-4

The inquiry into AI-powerd liturgy fits in well with a growing body of literature on technology and Christian worship. Having carried out a survey of interaction between technology and Christian worship in history, Susan White reminded us three decades ago that Christianity needs to confront technological challenges to Christian liturgy, and to reconsider the liturgy in technological culture in order not to marginalise Christian worship in technological society.[3] Some scholars examine the use of media technologies (e.g., video recording, live streaming) to enhance Christian worship, with the caveat that these technologies may debilitate traditional worship by breeding individualistic ownership of liturgy.[4] Others note that digital media bring up and store so much religious material as to give shape to online spirituality, which extends and yet somehow contradicts traditional offline worship.[5] Teresa Berger presents us with the hitherto most comprehensive analysis of interaction between digital technologies and liturgy, showing how established liturgical practices and notions may be transformed in the digial age.[6] I will have a closer look at some notions like virtual bodies and liturgical consciousness in the digital world. Despite much scholarship on technology and liturgy, what remains underexplored is how AI as a particular emerging technology may put exceptional pressure on Christian liturgy.

My objective in this chapter is to use Christian liturgy to provide a litmus test for the corollaries drawn in the previous chapters that the holistic reading of the *imago Dei* focuses emphasis on the idea of psychosomatic humanity and leads to the concept of AI as the ectype of psychosomatic humanity. Attention will particularly be drawn to the human biological body and mind, which are central to the meaning of being human vis-à-vis AI. While Christian litury will help us put the spotlight on factors that distinguish between humans and AI in the narration of God–human stories, these factors open up a way to develop AI-powered narration of God–human stories.

Following the principle of the ontological distinction yet connection between AI and humanity laid down in the previous chapters, it will be demonstrated in this chapter that AI-powered liturgical enactment can be considered the ectype of the human clerical enactment of liturgy for two reasons. Firstly, despite that it is an exaggeration to say that AI cannot be applied to liturgies due to the distinction between virtuality and reality, the uniqueness of human clerics in liturgical enactment can be defended by virtue of their psychosomatic humanity, more so when we attend to the role of their biological bodies (especially neural activities) in liturgical enactment. Secondly, human minds feature in liturgical enactment in such a sense that the clergy's consciousness of enactment cannot be reduced to computation and that their memory of liturgy is more than storage of liturgical data. These two aspects together underwrite the importance of holistic humanity in liturgical enactment and, meanwhile, characterise AI-powered liturgical enactment as subaltern and ectypal action, pointing to the way in which AI participates in the human performance of the redemptive stories of God and humans.

3.1 The Emergence of AI-Powered Liturgical Enactment

Liturgy is an ideal point of contact between Christianity and AI. This is so because the Christian liturgy is the core expression of Christianity. Philip Hefner puts it well, 'the liturgy ... [embodies] within the church that revelation of God's unfolding and God's purposes for the world which is manifest in Christ and to which the church is called to bear explicit testimony.'[7] As the central and first-order praxis of Christian churches, the liturgy is, of course, among the first matters coming on the scene while discussing interaction between Christianity and AI. One may ask: to what extent can AI faciliate or undermine liturgical praxis? This question drives us to explore how AI as the ectype of humanity (*imago homini*) may be linked to the performance of God–human stories.

We should recognise that Christian liturgy is subject to the gravitational pull of AI in the twenty-first century. Technology nowadays is radically different from that of the First Industrial Revolution (1760s). The technologies in the second half of the eighteenth century were mainly used to improve productivity and shorten work hours. In that century, technology's bearing on religious communities was its assistance in liberating humans from mechanical and unworthy labours so that humans could, as Friedrich Schleiermacher argued, 'have peace and leisure to contemplate the world in themselves' and freely join religious communities.[8] In respect of Christian liturgy, the technologies of the First Industrial Revolution allowed believers more time to participate in Christian worship. By contrast, the Fourth Industrial Revolution—which involved the creation of AI-powered technology—generates a radically different relationship between technology and Christian liturgy such that now church worship is being transformed by AI-powered technologies.

This new technology–liturgy relationship comes to the fore through AI's permeation into liturgy. One example is the first AI gospel artist J. C. created by the company Marquis to write and sing gospel songs.[9] Using AI algorithms, J. C. can create new songs by recognising and replicating the patterns of existing songs. To be sure, with the aid of AI systems, songs written for Christian worship can adhere strictly to some theological requirements. For example, Robin Parry maintains that songs for worship is nothing other than singing the Trinity. That is, these songs should convey the doctrine of Trinity and strike a balance in attributing glory to the three persons of the Trinity.[10] AI algorithms and processing system can, of course, be capable of using songs in its database to rapidly create new songs that comprise this 'Trinitarian syntax' (a phrase taken from Parry).

Another latest and more eye-catching example of AI-powered liturgy took place at St. Paul's Church in Nuremberg, Germany.[11] A 40-minute Lutheran service was presided and managed by ChatGPT which was personalised by four virtual avatars (two young men and two young women). The sermon, prayers, and music were created by ChatGPT and Jonas Simmerlein, a theologian and philosopher from the University of Vienna. Although some participants were not content with the service and considered it to be mechanised and emotionless, it is nevertheless true that AI has now challenged traditional liturgy in such a way so as to redefine who can enact the liturgy and how the liturgy can be performed.

These two examples, the St. Paul's Church case in particular, foreground a critical question to be addressed in this study on theology and AI: does AI *fulfill* human responsibilities to enact the liturgy on all levels in the near term? Unlike traditional liturgies only performed by congragations under the leadership of clergy, AI opens up possibilities to isolate the enactment of Christian liturgies from clergy. Of course, there are multiple ways in which AI can be applied to liturgy. In addition to the above two examples, AI makes it possible that liturgies can be enacted in the virtual world (e.g., the metaverse) either by AI ministers or AI-assisted liturgical performance (e.g., virtual worship environments). My main concern raised in this chapter is about whether AI can be a genuine enactor of liturgy as the same as human clergy and whether AI can, to a certain extent, replace humanity in liturgical enactment. Furthermore, given the AI-powered liturgy at St. Paul's Church, a derived question is concerning whether and how the church can deploy AI technology into Christian worship. In this connection, the holistic reading of the *imago Dei* and the archetypal–ectypal construal of AI as the *imago hominis* will show the uniqueness of human liturgical enactment vis-à-vis AI-powered one through connection among human consciousness, biological bodies, and Christian liturgies. We will circle back to this subject after spelling out Christian liturgy and its nature.

3.2 Christian Liturgy: Recapitulating the Story of God's Salvation

3.2.1 Christian Liturgy and Its Christological Nature

A definition of liturgy in general has been laid down above, but we need to go into details of Christian liturgy in order to flag up the characteristics of the role that clergy play in liturgical enactment. To this end, let us begin with the psalmist's exhortation:

> O sing to the Lord a new song;
> sing to the Lord, all the earth.
> Sing to the Lord, bless his name;
> tell of his salvation from day to day.
> Declare his glory among the nations,
> his marvelous works among all the peoples.
> (Ps 96:1-3; NRSV)

The psalmist brings to open the core meaning of Christian worship—that is, singing God's salvation. Characteristic of this liturgical psalm sung in the Old Testament is God's salvation for all creation. All creatures restored in God's salvation are, according to the psalmist, called to sing a *new* song to God. This liturgical emphasis on salvation points us to a doxological hope for the salvation of the whole creation in Jesus Christ (Rom. 8:22–25). Hence, this liturgical emphasis on salvation is eventually in tie with the history of God's salvation culminating in Jesus Christ. J. D. Crichton's lengthy elaboration on the link between liturgy and history of salvation is worth quoting here.

> [T]he Christian liturgy is patterned on the history of salvation. There is the original initiative of God, who throughout the ages offers man his love. The history is carried forward by the redeeming events of Christ the Son of God, who on the cross gives himself totally to his Father and for the salvation of mankind. He ascends to the Father and sends upon his Church the Holy Spirit so that his *ecclesia* may continue in time and space and by the power of the Holy Spirit his redeeming work. The Church looks on to the consummation of all things when Christ will hand all things to his Father so that he may be all in all things (1 Cor. 15.28). To the Father through the Son and in the Holy Spirit, is the underlying pattern of the history of salvation. So is it too of the liturgy.[12]

Crichton makes it clear that the liturgy and the history of salvation are, in essence, intertwined in that the Triune God is both the object of Christian worship and the maker of salvation. The Triune God's *opera ad extra* is revealed both in the liturgy and throughout the history of salvation.

What does it mean to say that the liturgy is patterned on the history of salvation? Crichton does not flesh out the meaning of 'patterned.' From the above block quote, it follows that the liturgy is Trinitarian as the same as the history of God's salvation. It is the Triune God who is at the centre of both the Christian liturgy and the history of God's salvation. Viewed in this light, the meaning of 'patterned' can be best paraphrased by 'recapitulation.' In this connection, Jean-Jacques von Allmen presents us with a groundbreaking study. He asserts that the Christian liturgy is the recapitulation of the history of salvation that is given in Jesus Christ. He adds that:

> *recapitulare* normally means more simply to "sum up" or "confirm" or "repeat", and in this sense the term is perfectly appropriate: the [liturgy] sums up and confirms ever afresh the process of saving history which has reached its culminating point in the intervention of Christ in human history, and through this summing-up and ever-repeated confirmation Christ pursues His saving work by the operation of the Holy Spirit.[13]

This liturgical recapitulation is, von Allmen makes it clear, always surrounding the Son of God who was incarnate in history. Christian liturgies are united in the recapitulation of the history of salvation centred in the incarnation.

That which is characteristic of the doctrine of incarnation must underlie each Christian liturgy, be it Lutheran, Reformed, Roman Catholic, or Christian more broadly. Examined in light of the doctrine of incarnation, discrepancies between human and AI-powered liturgical enactment can, therefore, be applied to various liturgies across different theological traditions.

The doctrine of incarnation lays emphasis on the hypostatic union of Christ's divine and human nature. This union is the exemplar of the communion between God and humans, which is part of the significance of the liturgy. Following this, what should not be disregarded is the role of humanity in liturgy. Having the Triune God as the object of worship may give one an impression that a proper liturgy rests with God alone and has nothing to do with humanity. Derived from this viewpoint is the conviction that *the natural* does not necessarily factor in a proper liturgy. As will be seen, this conviction backs up many approaches to AI-powered liturgy.

Such a de-human liturgy collides with the role that humanity plays in the doctrine of incarnation. But what is humanity in reference to the incarnation? The Chalcedonian formulation of the hypostatic union is mainly concerned with the consubstantiality of Christ's humanity with human nature in general. While Chalcedonian Christology lays a foundation for subsequent Christological development, we should concede that, as Ian McFarland points out, Christ's humanity in post-Chalcedonian Christology 'tends to be marginalized as a point of dogmatic interest' in that Christ's humanity is not fully 'integral to his identity' and 'overshadowed by the power of his divinity that God is revealed.'[14] Indeed, Jesus's humanity was highly regarded in the nineteenth century such that many modern theologians downplayed the distinctiveness of Christ's divinity.[15] However, such a modern Christological revision did injustice to the relationship between the divinity and the humanity of Jesus Christ. As such, the doctrine of incarnation should pay close heed to Christ's humanity *in order to* articulate a proper relationship between Christ's two natures. This incarnational connection between humanity and divinity should also feature in the Christian liturgy, the recapitulation of the story of God's salvation culminating in Jesus Christ.

Having grounded in the doctrine of incarnation, the Christian liturgy should underscore human elements while giving prominence to God as the object of worship. But what are the human elements of liturgy? Ola Sigurdson's celebrated study on the incarnation and humanity can help us here. He argues that 'a doctrine of incarnation that does not take into consideration the anthropological implications of the human mode of being-in-the-world will remain abstract.'[16] Relying upon phenomenology—which stress both how things appear to humans and how humans are conscious of things appearing in a particular way—Sigurdson suggests that 'the human mode of being-in-the-world' implies 'how Christ has been received in *concrete, human practices*.'[17] Needless to say, these human practices must give birth to inquiry into the human embodiment and the biological body, affirming the importance of the doctrine of incarnation for the understanding of the human person as a psychosomatic whole.

This incarnational line of reasoning about humanity underpins our understanding of the human elements of liturgy and tailors all Christian liturgies to the theological implications of the human being as a psychosomatic whole. On this point, we come full circle back to the holistic construal of the *imago Dei*, as elucidated in Chapter 1. The human being as the ectype of God means both that the *imago Dei* is the psychosomatic human person and that being human is eccentric, that is, being related to others. In particular, the human person's relationship with the Triune God perfectly demonstrates that the human being is the *imago Dei* and the ectype of the divine archetype. It suffices to say that the holistic reading of the *imago Dei* echoes the incarnational approach to liturgy, all the more so when we dwell on human factors in liturgical performance.

3.2.2 *Holistic Humanity in Christian Liturgy*

From the perspective of the incarnation, the Christian liturgy as the recapitulation of the history of salvation should showcase that the redeemed humanity in Christ responds to the grace of salvation in worshiping the Triune God so as to display the psychosomatic human being as the ectype and the *imago* of the divine object worshipped. Following this incarnational rationale, I will demonstrate in this section that the theological findings in Chapter 1 consolidate holistic humanity in Christian liturgy and that the latter substantiates the former. In so doing, the difference between the human being and AI shall be crystallised, and a potential for AI-powered liturgy can be actualised while preserving the unique role that humans have to play in liturgical enactment. In the two subsections that follow, I will successively discuss the liturgical significance of the human body and the human mind.

3.2.2.1 *Bodily Worship*

Derived from the doctrine of incarnation, the theological weight of the body in liturgy has many passages of the Bible on its side. In the Old Testament, particularly in the Book of Psalm, God's people worshipped Yahweh with their bodies (e.g., Ps. 84:1-2), such as mouths and hands (e.g., 1Kings 8:22; Ps. 47:1; 63:4; 134:2; 141:2), through bodily actions, such as dancing and playing various instruments (e.g., 1Ch. 15:16; Ps. 92:2; 150:1-6; Isa. 38:20). In the New Testament, Paul the apostle forthrightly exhausted Christians of the first century 'to present [their] *bodies* as a living sacrifice, holy and acceptable to God, which is [their] spiritual *worship*' (Rom. 12:1; emphasis added, NRSV). He also described the body as the Holy Spirit's temple (1Cor. 6:19), presenting us with the image of the place where Israel was called to worship. Given such a body-involved liturgy, one may pose a question: how does the human body involve in the liturgy performed by a psychosomatic human person?

Integral to the answer to this question is the body's cooperation with the spirit in liturgy. It is symbolism that is widely used by theologians from various theological traditions to flesh out the role of the body in performing the liturgy. In developing a Reformed liturgical system for nineteenth-century Dutch churches, Abraham Kuyper (1837–1920), the founding father of Dutch Neo-Calvinism, argued:

> Kneeling as such is nothing more than having your body assume a posture that symbolizes the soul bowing before the majesty of God. And this is not in the first place to make the body convey what lives in the soul, but rather to deepen and strengthen the action of the soul through the harmonious cooperation of body and soul.[18]

In his position, symbolism does not mean that the body merely serves as an instrument through which the human soul operates. Quite the contrary, the body intensifies the liturgical action of the soul with the visible symbols of bodily actions (e.g., kneeling). There is no hierarchical relationship between the body and the soul.

A similar construal of bodily symbolism related to liturgy can be found in the work of Edward Kilmartin (1923–1994), a Jesuit priest and liturgical theologian. While spelling out the theology of liturgy, he underwrites a holistic view of the human being: 'the spirit of the human person depends on the body for its contact with the world. The spirit does not turn by itself, in a second act, to others and things. On the contrary, it is by bodily contact that the human spirit is activated to turn itself toward the surrounding reality.'[19] The human spirit cannot operate on its own to interact with the world outside. Without the body, the spirit is dysfunctional. To this extent, the body is the 'real symbol' of the spirit, which is to say that the body as the signifying and the spirit as the signified are united.[20] As per Kilmartin's liturgical theology, this symbolic unity makes tangible what is hidden and mysterious in liturgy, 'the one mystery of the divine self-communication.'[21] Compared with Kuyper, Kilmartin furnishes a more clear-cut account of symbolism: the body as symbol serves as an epistemological approach to liturgy. More specifically, the body brings about symbolic perception, that is, 'the ability to penetrate beyond the superficial appearances of reality.'[22] Human bodily participation in liturgy opens up a way for participants to perceive Christ's presence and worship God 'in, with, and through Christ.'[23] In this way, the body cooperates with the spirit (soul) in perceiving God liturgically.

This liturgical symbolism, loaded with epistemological significance, further develops in Lutheran liturgical theologian Frank Senn's study on liturgy and the human embodiment. He articulates a theology of embodied liturgy and places emphasis on the role of the body in the worship of the psychosomatic human being. An underlying principle is his conviction that 'we have nothing else with which to worship God than our bodily selves.'[24] Behind Senn's enchantment with the body in liturgy is his commitment to the doctrine of incarnation: 'The manner in which the body expresses itself liturgically makes the essence of the liturgy bodily visible just as the essence of God for us became bodily visible in the Word made flesh in Jesus. What we are to see in Jesus is what the liturgy primarily celebrates: the paschal mystery of his death and resurrection.'[25] In distinction to both Kuyper and Kilmartin, Senn forthrightly justifies bodily symbolism by the doctrine of incarnation and moves on to demonstrate the way in which the body is used in liturgy. He illustrates, for example, that senses and postures are of significance for liturgy. 'The more the senses and postures are used in worship, the more we can individually and collectively connect with what is happening in the liturgy and … sense the presence of God.'[26] From this vantage point, it can be argued that the meaning of liturgy generated by human interaction with God in worship can be perceived by virtue of the fact that human body can *sense* God's presence.

This bodily sense of God's presence underscores Senn's liturgical understanding of the correlation of the body and the mind.

> I learned to respond to God's touch by worshipping God with my body; I learned to use all of my senses as well as bodily gestures and postures …. I found that crossing myself and bowing or kneeling gave me a more immediate sense of God's sense. I had to be *self-conscious* about what I was doing, and the gesture or posture reminded me of why I was doing it. I was worshipping God with my body and it brought meaning to *mind*.[27]

Two observations can be made in reference to this passage. First, the body and the mind work together to impel the whole human person to worship God. In particular, Senn's formulation of 'brought meaning to mind' implies some biological processes that have impacts on the mind. As will be demonstrated, the neural activity in the brain illuminates a clear distinction between human and AI-powered enactment of liturgy. Second, all of human senses, bodily gestures, and postures used in liturgy remind liturgical participants that they are worshipping with other persons *in that space* and *at that time*. Specifically, sense perception is indexed to the spatiotemporal factors of liturgy, and gestures and postures are often uniform for corporate worship. We will come back to go into details of liturgical spatiotemporality in Section 3.4. Now we are moving on to briefly examine the human body in corporate worship.

In the recapitulation of the history of God's salvation, human participants are conscious of their co-worshippers—whose human natures are redeemed in Christ, the centre of the liturgy—through senses, such as sight, hearing, and touch. Carl Braaten and LaVonne Braaten put it well: 'In the body we are individualized and personalized. Yet no one is wholly and only his body. He is always a member of a larger body together with others, as long as he is a living body.'[28] The corporate liturgy offers an occasion when the church as the corporate body is evident in its unity of individual bodies. This bodily corporate worship helps us to zoom in on the meaning of the communion of saints (*sanctorum communionem*). Aside from the common construal that the communion of saints refers to the spiritual union between the faithful on earth and throughout history, theological emphasis on body-involved liturgy connotes that such a communion is also biological and tangible. Both corporate and individual bodily worship belong together in liturgical symbolism that underlines the way in which the body brings the meaning of liturgy into the mind.

3.2.2.2 Conscious Worship

Christian liturgy patterned on the doctrine of incarnation should give attention to humanity in its full sense, that is, the human being as psychosomatic unity. From this it follows that liturgy is as much body-involved as mentally committed, which means that the body and the mind together play a liturgical symphony. Having spelled out bodily worship, we move on to inquire into the following question: how does the human mind play a part in liturgy? As has been noted above, the meaning of liturgy cannot turn up in the mind without the body. Therefore, what remains to be explored is how the mind responds to sensory stimuli through the body during liturgy.

The meaning of liturgy is generated through human interaction with both God and co-worshippers. I have touched upon this point above. Recent studies on the application of embodied cognitive science to religion can help us delve into how the mind operates in liturgy. The premise of embodied cognition aligns with the principle of holistic anthropology in this study—that is, the human being as a psychosomatic whole. Correlating embodied cognition and religion, Fraser Watts argues that '[e]mbodied cognition is certainly physicalist in its assumptions in the broad sense of emphasizing the physical context of cognition.'[29] There are two observations to be made from Watts's assertion: (1) humans are embedded within physical contexts by virtue of their physical bodies; and (2) these physical contexts trigger cognitive effects, facilitating the formation of human thinking, understanding, consciousness, and other mental states. Watts suggests that the liturgy as a religious and physical context encloses our physical bodies and makes our liturgical postures, gestures, and sacraments meaningful and cognitive for the mind.[30]

Simulation—taking place in the mind during cognition—'involves the reenactment of perception, action, and even interior states resulting from our embodied experience with the world and ourselves,' being linked up both with neural activities and nonneural structures.[31] Viewed in this light, the human consciousness of liturgy depends both on the integration of human social interaction within the liturgical context and on what takes place *in* the human body (including neural activities) and mind. In this way, the liturgy *per se* is constitutive of the conscious states of the mind. An example of evidence for this corollary is Robert Fuller and Derek Montgomery's experiment about body postures and religious attitudes. Through the experiment they observe that agreement with and attitudes towards religious beliefs are related to body postures. That is, 'lower, contractive positions' result in more committed religious beliefs.[32] In other words, lower postures performed in Christian liturgy facilitate one's *knowing* of the history of God's salvation. Neural activities have a role to play in one's mental transformation happening during liturgy. We will return to discussion about neural activities and liturgy while contrasting the human enactment of liturgy with the AI-powered one. At this point, my emphasis will fall on the memory of liturgy, a crucial aspect of the human consciousness of worship.

Memory has caught much scholarly attention in the fields of both cognitive science and the philosophy of mind. The human mind and consciousness would not exist without memory, argue Richard Thompson and Stephen Madigan.[33] As such, memory is placed at the core of conscious worship. The Anglican liturgical theologian Peter Atkins presents us with a path-breaking study on memory and liturgy.

> *At the heart of all worship is the act of remembrance* .... In Christian worship the object of remembrance is the God who is revealed in Jesus Christ and known through the power of the Holy Spirit. Each occasion of worship allows the worshipper to recall *consciously* the relationship formed with Christ, and this recollection forms links with all the previously stored experience that the worshipper has of the presence of the Spirit of Christ in the past.[34]

From Atkins's sentiment we can make three inferences: (1) the memory as a prerequisite for liturgy; (2) the unity of the past, the present, and the future in the memory of liturgy; (3) mutual strengthening of consciousness and the memory in liturgy.

First, the memory is a *sine qua non* for liturgy in that the knowledge of God's salvation must have been stored in the worshipper's memory while recapitulating the history of God's salvation. Worshippers should know the object of worship before liturgy, which means that this object is already in their memories as well as consciousness. In theological terms, this already-existing knowledge of God as the object of worship is the consequence of salvation wrought by God in Christ. Every liturgy provides worshippers with an occasion to retrieve their knowledge of God from their memories and correlates it with the here-and-now liturgy. An excellent example of this retrieval is the Eucharist. Each Lord's Supper is, to borrow Atkins's words, 'remembering Jesus Christ'—that is, recalling God's salvation in Christ through the bread and the wine. In this vein, Christ is *really* present in the liturgy of the Eucharist. 'Because [Christ] had related bread to his body and wine to the blood of his sacrifice, his disciples could remember his full self and what he had done for humanity.'[35] All in all, the liturgy is to recapitulate the memorised salvation of God.

The second inference is that the worshipper not only recalls what is remembered in the past, but also remembers what is performed here and now and looks to what is to come. Through retrieval of the knowledge of God in memory, worshippers link previous stored liturgical experience to the present liturgy and look to what is promised in liturgy. In Atkins's position, this unity of the past, the present, and the future rests with God himself. 'Worship is focused on the God who was, and is, and will be. In worship we draw on our experience of God in the past at the same time as we look for a new experience of God in the present. Our worship prepares us for our expectation of God in the future, and for our service of action in God's name.'[36] By virtue of God's omnipresence from past to future, Atkins is convinced that human liturgical memories will grow all along and reach to what is yet to arrive. Having in mind liturgy as the recapitulation of the history of God's salvation, such *remembering the future* during liturgy is tied up with the consummation of God's salvation in Christ. It is in liturgy that a distinctive feature of the church comes to be seen: 'The Church stands within the dynamics of memory and hope, grounded in the past events that shape her life—Christ's life, death and resurrection—and awaiting the fulfilment of God's Kingdom.'[37] As such, a bold claim can be made that, unlike normal memories, liturgical memories are teleological and capable of ceaselessly giving birth to future memories directed towards the unexpected end. (In Section 3.2, I will flesh out this point in contrast with liturgical data stored for AI.)

Atkin's sentiment on memory and liturgy yields a third inference that worshippers's consciousness of liturgy reinforces their liturgical anamnesis, and their memory of liturgy, in turn, fosters their consciousness of worship. From this vantage point, we can discover that each occasion of worship involves double mental operations. Firstly, reinforcing anamnesis means that there may be a loss of memory without liturgy. According to Atkin, '[o]ne of the key reasons for worship is to prevent spiritual amnesia—the loss of memory as to who we are and who God is.'[38] Worshippers's knowledge of both their identity and God is refreshed and consolidated in liturgy. Coupled with the second inference, it can be argued that the unity of the past, the present, and the future in memory takes shape in here-and-now worship, which means that the consciousness of the immediate liturgy amplifies worshippers's religious anamnesis.[39] Secondly, worshippers's memory of liturgy brings up their consciousness of here-and-now worship. Their conscious relationship with God in liturgy is the very relationship that they have remembered. As memory is central to consciousness, liturgical memory and consciousness simultaneously operate on each other.

The worshipper's consciousness and memory of liturgy together signals something more important than an individual worshipper. Granted, liturgical consciousness and memory often refer to the mental state of one human worshipper. However, a worshipping human person's memory of liturgy is always concomitant with that person's consciousness and remembrance of fellow worshippers *around* them at liturgical occasions. Joshua Cockayne and Gideon Salter strike a right note that memory and belonging come together.

> [R]emembering cannot be separated from belonging. Whether this is a matter of the propositional knowledge we are able to recall, the shaping and re-shaping of a communal narrative through repeated reminiscing or the pre-reflective enactment of a family's 'way of being', we remember in order to identify with a community. It is through memory that communities are able to persist.[40]

In this light, communal memories develop in liturgy, and communal identities take shape and are strengthened in such memories. Furthermore, from the interplay between the individual worshipper's memory and the worshipping community's memory we can infer the communal implication of the second inference made above—the unity of the past, the present, and the future in liturgy. Specifically, as Atkins suggests, the liturgy's 'story, songs, teaching and ritual pick up the memory of God's people in the past and allow it to empower and inform them for the present and the future.'[41] This observation goes some way to illuminating that the liturgical remembering of what is yet to come by virtue of salvation takes place in worshipping communities.

To sum up, the idea of conscious worshipper shows that a worshipping human person participates in the liturgy as much with the mind as with the body. In particular, the human memory of liturgy is intertwined with the role that consciousness plays in worship. Human worshippers are conscious of liturgies, which were in their memory, are being remembered individually and communally, and point to something to be remembered.

3.2.3 The Clerical Enactment of Liturgy: An Action of the Holistic Human Being

In the preceding subsection, I have demonstrated that the incarnational foundation of liturgy brings forth the notion of holistic human worshipper. My emphasis has thus far fallen on human worshippers in general—that is, how humans *participate in* liturgy both with bodies and minds. That is to say, the context for the above discussion is a liturgy that has already been enacted by clergy. Hence, what remains to be examined is how the notion of holistic humanity in liturgy is related to the clerical enactment of liturgy, the subject matter of this chapter.

While embarking on dealing with this question, a tenet should be laid down: the clergy, who enact the liturgy for participants, are co-worshippers with congregations. Derived from this tenet is the objection to a false belief that while clerics are liturgical enactors, congregations are simply audience or observers. Nicholas Wolterstorff's analysis of the nature of the clerical enactment of liturgy can undergird this tenet.

> [T]he liturgy is not something enacted by the clergy for the purpose of satisfying the needs and desires of the congregants, be those needs and desires spiritual, emotional, aesthetic, or whatever. It's the church that enacts the liturgy, not the clergy. Though the church does so under the leadership of the clergy, the liturgy is not something that clerics do. And the church enacts the liturgy not to satisfy the needs and desires of individual congregants but to worship God. ...It's not the individual members who do these things simultaneously; it's the assembled body that does these things.[42]

In Wolterstorff's position, the clergy's enactment of liturgy should be construed in a broader, communal context. Rather than being the cleric's individual action, liturgical enactment is a communal action that only aims at the worship of God. This communal nature of liturgical enactment strongly echoes Joseph Ratzinger's reminder that worshippers should 'look together toward the Lord' rather than the clergy who enact the liturgy.[43] All the more important is that this communal nature sheds light on the fact that clerics and other worshippers belong together, which ensures that clergy's solidarity with congregations is prioritised over their divergence in terms of liturgical enactment. Following this we can infer that a cleric should be a bodily and conscious worshipper before taking the role of an enactor of liturgy. What we discussed in the above subsection 3.2.2 'Holistic Humanity in Christian Liturgy' lays at the foundation of liturgical enactment. As will be fleshed out in Section 3.3, these psychosomatic features of human worshippers will elucidate the extent to which human and AI-powered liturgical enactment is connected yet distinguished.

That said, this communal nature of liturgical enactment does not rule out the particular *task* of clerics in liturgical enactment. The term 'task' is deliberately employed to convey the pragmatic, functional, representative significance of the clergy's action in liturgical enactment. This triple significance is rooted in the nature of liturgical enactment, which is, as Wolterstorff observes, 'the site of mutual address and listening between God and the people.'[44] It is in liturgy that God and humans address and listen to each other. Mutual listening between God and humans indicates human relationship with God. That is, there is 'an I-thou relationship between God and us. God is a thou for us.'[45]

While discussing God's address to humans, Wolterstorff draws on von Allmen's view of the three modes of God's address in liturgy. The first mode is Scripture-reading, which can be performed by congregations. According to von Allmen, Scripture-reading is reflective of the incarnational nature of liturgy: 'to read Scripture is to experience the paschal joy; the Lord reappears, He who is the Word, to tell us of His love and His will to teach us who He is and who we are, to summon us and give us life.'[46] In distinction to the first mode, the other two modes of God's address are mainly related to clerics. That is, God's address to worshipping humans takes place both through the proclamation of God's Word and at particular liturgical moments. God's address to humans through preaching 'a direct prophetic intervention in the life of the faith and of the Church.'[47] In addition, at some particular moments during liturgy, a cleric 'declares and gives to the people the greeting, the absolution and the blessing of the Lord.'[48] These two modes of God's address to humans bring to light the particular task of clerics in liturgy. The clergy's enactment of liturgy is pragmatic insofar as both their proclamation of God's Word and other speaking to worshippers pertain to liturgies taking place within particular communities here-and-now. Clerical enactment is functional insofar as clerics assume particular roles in delivering God's address to worshippers in liturgy.[49] The clergy's action in liturgy is representative precisely because they simply represent God who is the authentic speaker addressing to all worshippers.

The clergy's particular task in liturgical enactment should be understood in conjunction with both the communal nature of liturgical enactment and the solidarity between clerics and other worshippers. From this we can infer that the bodily and mental nature of worship—which is characteristic of the communal enactment of liturgy—should feature in the particular task of clerics in liturgy. The triple significance of the clergy's action in liturgical enactment is linked up with the functioning of the clergy's bodies and minds. This assertion will occupy the central place in the following discussion in Section 3.3 over the clergy's liturgical enactment vis-à-vis the AI-powered enactment.

3.2.4 *Summary*

Through the theological lens of the incarnation (the centre of the history of God's salvation), it has been demonstrated that Christian liturgy is a bodily and mental action in that the bodies and minds of psychosomatic worshippers work together to participate in liturgy. This *holistically anthropological feature of liturgy* aligns with the theological idea that the term *imago Dei* points to the stories of God and *holistic human beings*. The symbolic implications of the senses, postures, and gestures of spatiotemporal liturgy construed within bodily worshipping communities turns our attention to the connection between liturgical symbolism and the mind. Worshippers are conscious of remembered and to-be-remembered liturgies through both bodily interaction with liturgical contexts and biological particularities. This psychosomatic feature of worship is common to all humans participating in communal worship, a liturgy enacted by the whole church. As such, the clerical enactment of liturgy should be psychosomatic, and both the clergy's bodies and minds play an equally significant part in enacting liturgies pragmatically, functionally, and representatively. In the section that follows, I will unfold the psychosomatic properties of the clergy's liturgical enactment through exploration of AI-powered liturgical enactment.

3.3 Enhance or Weaken: AI-Powered Enactment of Liturgy

Having demonstrated Christian liturgy as recapitulating the history of God's salvation in a psychosomatic way, a question that plagues us arises: does AI-powered liturgical enactment fit in with such psychosomatic recapitulation? The answer to this question should expand on whether AI enhances or weakens liturgical enactment. In correspondence to the above subsection 3.2.2, what will be elucidated below consists mainly of two parts. The subsection 3.3. 1 will examine the idea of bodily enactment through exploration of how human biological traits distinguish between human and AI-powered liturgical enactment. The subsection 3.3.2 will probe the role of the human mind in liturgical enactment with an emphasis on the differences both between artificial and human consciousness and between human memory and AI's data storage. In the course of the following discussion, the main claim made in the previous two chapters will be reiterated: AI is the ectype of the *imago Dei*, differing from yet mirroring humanity in liturgical enactment.

3.3.1 Bodily Enactment

AI-powered liturgical enactment for churches can come into effect in many ways. One instance is the Avatar-assisted ChatGPT at St. Paul's Church in Nuremberg, Germany, as has been noted in the Section 3.1. Throughout the service, four virtual avatars enact the liturgy and orient all participants during this course. In this way, AI-powered liturgical enactment takes place in a brick-and-mortar church buildding. A second instance is the church in the metaverse, such as Life.Church.[50] Not only do worshippers participate in liturgy in the digital way of avatar, but the liturgy can simply be enacted by AI. In this case, AI-powered liturgical enactment comes about in the virtual world. Despite the difference between the applications of AI to liturgical enactment in different contexts, all AI-powered liturgical enactments give rise to the same question about holistic humanity in liturgy: What role can the clergy's biological bodies play in the liturgy provided that liturgies in their minds are programmed into AI systems?

3.3.1.1 Not the Antithesis between Virtuality and Reality

While considering the role of the clergy's bodies in liturgical enactment, a major concern has been raised about the contradistinction between virtuality and reality in present literature on liturgy and technology. In his recent studies, William Young alerts us to the challenges raised by autonomous systems and AI technology—such as virtual pastors, digital clergy, and Rev. Robot—to faith communities, including, *inter alia*, virtual pastoral care, virtual liturgy, and online spirituality.[51]

Teresa Berger presents us with more fine-grained and extensive inquiry into the question of virtuality in relation to liturgy. It is worth noting that her inquiry into digitally mediated liturgies does not touch on AI. Yet, her appreciative appraisal of digitally mediated liturgies is heuristic and helps us deal with issues surrounding AI-powered liturgical enactment. A rationale behind her research is that we should 'interpret digitally mediated liturgical practices within a larger trajectory of mediated practices in worship across time. [We] also understand these practices as part of larger contemporary cultural transformations of which they are a part.'[52] This rationale lays a foundation for her analysis of the potential virtuality of liturgy. Berger contests that it is misleading to consider digitally mediated liturgy virtual or un-real.

> The notion of liturgical life in cyberspace can arouse substantial misgivings. These are typically rooted in the assumption that [digitally mediated] worship amounts to a dis-embodied, virtual, and therefore un-real practice. The truth behind these misgivings is the fact that human bodies are the basic materiality and prime media of Christian worship. In contradistinction, digitally mediated practices seem to foster an illusion. They are virtual, after all, rather than real.[53]

Berger registers that underlying these misgivings is an unconvincing, 'inadequate and outdated' presupposition that 'virtual' is an antonym of 'real.'[54] She also takes issue with the identification between the virtual and the disembodied because human bodies and bodily actions are necessary for the entry into the digital world.[55] Taking these claims together, she contends that '[d]igitally mediated liturgical practices are material practices, as are all offline liturgies.'[56]

Berger's objection to the distinction between reality and virtuality comes to the point. Philip Brey reminds us that the ontological implication of virtuality puts an end to the separation between the virtual and the real. For all that the virtual is often considered fictional, we should not be oblivious to the fact that the virtual can interact with us and 'may stand in causal relationships to other entities.' For this reason, the virtual cannot simply be considered fictitious or unreal.[57] With the rapid progress of digital technologies, AI *par excellence*, the connection between the real and the virtual has been becoming more solid. Luciano Floridi puts it well: 'We have moved inside the infosphere .... [O]ur understanding and conceptualization of the essence and fabric of reality is changing. Indeed, we have begun to accept the virtual as partly real and the real as partly virtual.'[58] As such, the intertwinement of the real and the virtual is in lieu of the antithesis between them. From this follows the necessity of ontology of virtual reality, as Massimo Durante claims.[59]

> Digital technology generates the new ontology of the virtual. It is pointless and perhaps even counterproductive to try and draw a line between physical and virtual reality in strict ontological terms, for it is we—through technological innovation—who move and reshape that line. It is thus more relevant and fruitful to ask in pragmatic terms how the technology of the virtual modifies the way we experience reality.[60]

It is clear that Durante's contestation sits well with Berger's appreciative appraisal of digitally mediated liturgy, which is what exactly happens with AI-powered liturgical enactment.

An anticipated objection to AI-powered liturgical enactment is that such AI clerics are merely virtual and have nothing to do with the *real* enactment of liturgy. Neither AI-powered liturgical enactment in brick-and-mortar church buildings nor that in the purely virtual world is real and physical. However, the contestations raised by Berger, Brey, Luciano, and Durante convince us that there is *the intrinsic connection* between the real and the virtual. From this vantage point we can see that AI-powered liturgical enactment should not simply qualify as virtual in the nonphysical or unreal sense such that its reality is obliterated. But how shall we construe such an intrinsic connection? What is the ontological implication of such an intrinsic connection, echoing Durante's appeal for the ontology of virtual reality?

The idea of AI as the ectype of humanity is a conceptual apparatus through which to theologically elucidate the ontological implication of AI-powered liturgical enactment. In so doing, it will come to the fore that there is no rivalry between the virtual and the real in the case of AI-powered liturgical enactment. We need to recall the three implications of AI as the ectype of humanity demonstrated in Chapter 1, Section 1.4: (1) AI is created in the image of the *imago Dei*; (2) AI is eligible to participate in the performance of God–human stories by virtue of the close tie between AI and humans, though such participation cannot be actualised on AI's own; (3) AI as the ectype of the human being should reflect the moral nature of the *imago Dei*. The third implication will be examined in the chapter that follows. The first two implications are of relevance here.

The first implication brings up the ontological link established by the idea of *imago Dei* between AI-powered liturgical enactment and AI as the ectype of humanity. I have pointed out at the outset of this chapter that the *imago Dei* signifies human communion with God in worship. William Dyrness articulates a cogent account of the ontological importance of the *imago Dei* for worship.

> The story that shapes worship includes the assertion that human beings are created in the image of God, created to respond to and, ultimately, to worship the Creator. The end for which humans are made includes

> centrally the call to respond to God in prayer, praise, and thanksgiving. The Christian story contends that we are most fully human when we stand in the presence of God and give thanks.[61]

Since AI is created in the *imago hominis*, a mirrored role in liturgy should be endowed to AI such that *the end for which AI is created* includes its participation into the performance of the God–human story. To construe the ontological implication of AI-powered liturgical enactment in light of AI as the ectype of humanity, the intrinsic connection between AI and humans in liturgical enactment come to be seen. It is myopic to say, therefore, that avatar-assisted AI ministers as the *imago hominis* are not eligible to enact liturgies due to their virtuality and non-physicality.

That said, AI-powered liturgical enactment is not independent of humans because of AI's ectypal nature, as the second implication of AI as the ectype of humanity indicates. AI-powered liturgical enactment is automated and not an action personally performed by the *imago Dei*. While attending to the connection between the *imago Dei* and the *imago hominis*, we should not leave out the ontological distinction between the human archetype and the artificial ectype. This ontological implication of AI-powered liturgical enactment prompts an inquiry into the following question: How do virtuality and reality distinguish between yet connect with each other in terms of AI ministers?

As the ectype of humanity, AI should somehow display the feature of *imago Dei*. I have spelled out in Sections 1.1 and 1.2 of Chapter 1 that the holistic reading of the *imago Dei* is comprised of two aspects: (1) the psychosomatic human being as the *imago Dei*, and (2) the collective meaning of being human. The second aspect sheds light on the potential traits of AI ministers. That is, the ectypal feature of AI as the *imago hominis* implies its collectiveness, which means that AI-powered liturgical enactment should be understood in its social relation to worshipping communities.[62] To say the least, AI-powered liturgical enactment presupposes that there will be worshippers to participate into the liturgy. AI ministers *per se* are embedded with a social purpose, which springs from their human creators who are the *imago Dei*. From this we can make two inferences. Firstly, the *virtual* cleric enacts the *real* liturgy within the worshipping community, and the virtual and the real belong into the same community. Secondly, the sociality of AI-powered liturgical enactment should be construed in view of the role that the human body plays in the *spatiotemporal* worshipping community. Here we return to the first aspect of the holistic reading of the *imago Dei*: the *psychosomatic* human being as the *imago Dei* ensures that human sociality is intertwined with spatiotemporal human bodies. Following this we arrive at a corollary that although the virtual is not contrary to the real in the case of AI-powered liturgical enactment, there is another contrast related to the body in liturgical enactment: the silicon-based versus the biological.

3.3.1.2 But the Contrast between the Silicon-based and the Biological

While considering AI ministers in relation to liturgy, we cannot simply pit the virtual against the real. The archetypal–ectypal ontology of AI underpins the *virtual reality* of AI-powered liturgical enactment. The emphasis on the human body and physical embodiment cannot lead to the claim that nonphysical or non-bodily liturgical enactment should be declined altogether. With that being said, we should ward off a naïve belief that AI-powered liturgical enactment can be uncritically received within worshipping communities. At this point, attention must be given to the distinction between human biological bodies and AI's silicon-based systems. I have discussed how the distinction between the silicon-based and the biological is related to the idea of AI as the ectype of humanity (Chapter 1, Section 1.4). Here I seek to flesh out how this distinction consolidates the holistic reading of the *imago Dei* as developed in this study. Since biological particularities are part of the meaning of being human, the difference between silicon-based AI and human biological bodies should be underscored while examining the application of AI to liturgical enactment.

We need to recall the discussion in the above subsection 3.2.2.1 over the importance of human bodies for worship. A crucial observation made there is concerning the worshipping biological body's participation into liturgy and communion with co-worshippers through human senses, gestures, and postures, which brings the meaning of liturgy into the mind. I also spelled out in the above subsection 3.2.3 that inasmuch as clerics are part of the worshipping community, biological bodies should feature in their liturgical enactment. A question then arises: Why is the contrast between the silicon-based and the biological often overshadowed by that between the virtual and the real in relevant debates about digitally mediated liturgy?

A major cause of such overshadowing is the attempt made to reduce organism into computation. In Chapter 2, I have spent some space examining the computational model of consciousness in the subsection 2.1.2 and elaborating on the connection between consciousness and the brain in the subsection 2.2.2. A further step needs to be taken to explore humans *qua* organisms vis-à-vis computational technology. This inquiry is particularly relevant to Yuval Noah Harari's idea of algorithmic organism. He defines algorithm as 'a methodical set of steps that can be used to make calculations, resolve problems and reach decisions. An algorithm isn't a particular calculation, but the method followed when making the calculation.'[63] In his position, an organism's life, work, and action are controlled algorithms, a far more complicated methodical set of steps than those algorithms applied to machines.[64] The same goes for humans. There are, so to speak, biological algorithms within humans, and human imagination, intelligence, emotions, and memories can be biologically explained as the consequences of algorithmic operations in humans.[65] From this it follows that human biological particularities can be replicated in a silicon-based systems (e.g., AI systems) in the algorithmic way. As such, what is at issue in AI-powered liturgical enactment is less concerning whether the silicon-based system can substitute for the biological body than concerning the contrast between the real and the virtual. Whether the liturgy is enacted by human biological hands or avatar-assisted AI ministers in silicon-based systems is no longer a perplexing question.

Antonio Damasio takes issue with Harari's identification between organism and algorithm. Damasio argues that the algorithmic account of organism downplays the value of biological substrates on which organic algorithms operate.[66] All the more important is that algorithms fail to epitomise the living conditions of organisms. Damasio's lengthy elaboration on the differences between algorithms and living organisms is worth quoting here.

> Saying that living organisms are algorithms is in the very least misleading and in strict terms false. Algorithms are formulas, recipes, enumerations of steps in the construction of a particular result. Living organisms, including human organisms, are constructed according to algorithms and use algorithms to operate their genetic machinery. But they are not algorithms themselves. Living organisms are consequences of the engagement of algorithms and exhibit properties that might or *might not* have been specified in the algorithms that directed their construction. Most important, living organisms are collections of *tissues, organs, and systems within which every component cell is a vulnerable living entity made of proteins, lipids, and sugars*. They are not lines of code; they are palpable stuff.[67]

Damasio makes it clear that while algorithms go some way to characterising living organisms, their biological components and particularities cannot be reduced to algorithms. The biological substrates of human bodies play a significant part in the human being that is integrated as a living organism.[68] From this vantage point we can see that the clergy's biological bodies operate in a distinct way in enacting liturgies vis-à-vis silicon-based AI. In other words, the uniqueness of the clerical enactment of liturgy, compared to AI-powered liturgical enactment, is partly yet significantly underpinned by human biological conditions.

Simon Balle and Ulrik Nissen's recent study on human bodies versus AI and robots helps to depict the distinctive role of human biological bodies in liturgical enactment. They assert that human biological bodies serve as remarkable boundaries between humans and AI.

> And, because we are biological beings, we are embedded in the world in a certain way that at every giving moment makes certain parts of it relevant for us to thrive, biologically, psychologically, socially. Our relation to the world is shaped by our organismic needs to interact with it to survive and thrive. In other words, our knowing the world is shaped in countless prereflexive ways, both by what we can sense and how our motor functions allow us to interact with the world, but just as fundamentally by our biological, social, and emotional need to understand it.[69]

In Balle and Nissen's position, human biological particularities give shape to a way in which humans interact with everything surrounding themselves. Such interaction reflects the distinct responsivity of human bodies. With biological bodies, humans are 'sensitive and responsive to changing conditions' so as to '[move] on their own accord toward more favorable ones in which to flourish.'[70] AI lacks this biologically derived responsivity and, consequently, fails to be responsive to changing, new situations.[71]

Recalling the aforementioned notion of the liturgy as the site of mutual address between God and worshippers, it can be argued that the human body's responsivity showcases its special function in liturgical enactment, which cannot be emulated in AI's silicon-based system. Worshippers together are responsive to liturgy through addressing to God. Belonging into the worshipping community, the clergy should also be responsive to liturgy while enacting the liturgy. Here we need to reiterate the two modes of God's address to worshippers through the pragmatic, functional, representative task of clerics: the proclamation of God's Word, and the declaration of God's greeting, the absolution, and the blessing at particular liturgical moments. Both the proclamation and the declaration presuppose that clerics *are capable of being responsive to* liturgies that may vary across circumstances and time. Moreover, the actualisation of one cleric's responsivity varies at different occasions of liturgical enactment. The clergy's enactive actions, including the administration of sacraments, are not mechanically repeated at each worshipping occasion. Rather, the clergy's each liturgical enactment is their new response to a new liturgical situation through performing God's address to worshippers. Responsivity is necessary for liturgical enactment in changing, new situations.

Human and AI-powered liturgical enactment radically differs due to the essential difference between the silicon-based system and the biological body. The different material causes of clerical and AI-powered liturgical enactment matter to worshippers. Granted, the idea of AI as the ectype of humanity lays an ontological foundation for the virtual reality of AI-powered liturgical enactment. Yet, this ontological implication does not underwrite the substitution of AI ministers for clerics in liturgical enactment. Rather, the contrast between the silicon-based system and the biological body in liturgical enactment strikes a chord with another implication of AI as the ectype of humanity: AI as the *imago hominis* does not emulate the *imago Dei* that is in the psychosomatic unity. Both the mental and the biological aspects of humanity factor in the clerical enactment of liturgy.

3.3.2 *Conscious Enactment*

Having elaborated on the unique role of the cleric's body in liturgical enactment vis-à-vis the silicon-based system, I move on to probe the meaning of the cleric's *conscious* liturgical enactment versus AI-powered liturgical enactment. It has been extensively demonstrated in Chapter 2 that if AI has consciousness in a broader sense of this term, then artificial consciousness can be considered as the ectype of creature consciousness. In what follows, I will flesh out how this ectype of creature consciousness underlies AI-powered liturgical enactment. Two interrelated contrasts will be investigated: (1) the *conscious* clerical enactment of liturgy versus the *computational* AI-powered enactment of liturgy; (2) liturgical enactment *remembered* by clerics versus liturgical enactment *stored* in AI. By doing so, I aim to bring to light the unique role of the clergy's minds in liturgical enactment and, at the same time, demonstrate the extent to which AI-powered liturgical enactment is connected to yet different from the clerical enactment of liturgy.

3.3.2.1 *Because of the Contrast between the Conscious and the Computational*

We need to invoke two principles elucidated earlier: (1) the worshipper as a holistic human person consciously participates in liturgy; (2) belonging into the worshipping community, the clergy enact the liturgy as consciously as they worship. With these two principles in mind and coupled with the findings of Chapter 2, I seek to uncover why AI-powered liturgical enactment should be considered computational. To this end, I shall restrict my attention to two aspects of conscious liturgical enactment: (1) the semantic nature of conscious liturgical enactment, and (2) conscious liturgical enactment as an intentional action.

It has been reiterated that worship is tied up with the knowledge of God's salvation because the liturgy is patterned on the history of salvation. In liturgy, worshippers will recall, renew, and increase their knowledge of God, who is the unique object of worship. In liturgy, a crucial function of the worshipper's mind is her understanding. On each occasion of liturgy, the worshipper must *know* and *understand* the meaning of liturgy. By the same token, clerics should *know* and *understand* liturgies, more so when enacting liturgies.

Nicolas Wolterstorff's analysis of scripted liturgy underwrites the important role of understanding in the clerical enactment of liturgy. He suggests that all liturgies—be it Reformed, Lutheran, Roman Catholic, Eastern Orthodox, the Evangelical, or Christian more broadly—are enacted in a similar way as performing musical works—that is, being performed according to *certain rules*.[72] He designates 'the complete set of rules for a correct liturgical enactment, both those specified in the liturgical text and those to be found in the relevant liturgical culture, the script for the liturgy.'[73] As such, the liturgy is a scripted and rule-governed activity.

Wolterstorff maintains that scripted activity is indispensable for the lives of conscious human beings. Throughout their lives, all human beings need to understand numerous prescriptions in order to know how they should perform required and permitted actions.[74] For a scripted activity, a liturgical script 'specifies a sequence of actions to be performed. Those actions are not just the utterance of certain words or the making of certain gestures. They are the uttering of certain words and *saying something thereby*, or the making of certain gestures and *doing something else thereby*.'[75] As such, he calls the liturgy the 'know-how', which means 'knowing how and when to perform the scripted actions' in liturgy.[76] Furthermore, this liturgical know-how is communal in character, which means that it is shared among the worshipping community and handed on from generations to generations, thus giving shape to different liturgical traditions.[77]

Given their particular task in liturgical enactment, clerics who share liturgical know-how with co-worshippers should comprehensively and thoroughly know and understand the script of liturgy, more so at particular moments when they address to the worshipping community while preaching and declaring God's greeting, absolution and blessing. It is at this point that a fundamental difference between human and AI-powered liturgical enactment comes to the fore. Liturgy as scripted activity requires clerics to follow specific rules consciously while enacting liturgy. This rule-governed nature of liturgy raises the question: Why does AI that follows certain rules created through algorithms differ essentially from human clerics who follow the liturgical script in terms of liturgical enactment?

To address this question, the concept of rule holds significance. Lorraine Daston's extensive inquiry into the meanings of rule can help to unravel the mysteries of rules in relation to the AI-powered and clerical enactment of liturgy. In her Laurence Stone Lectures, Daston classifies rules into three categories: rules-as-algorithms, rules-as-laws, rules-as-models or- paradigms. Rules-as-laws are rigidly and uniformly formulated and aspire to be applied universally.[78] Rules-as-models refer to examples and paradigms that point beyond themselves, and so they are to be emulated and followed with judicious discretion. These rules are thickly formulated—ranging over wide variations—and are flexibly applied.[79] Rules-as-algorithms are relatively thinly formulated, which target foreseeable possibilities and do not take unusual or diverse cases into consideration. Hence, their application does not allow for the exercise of discretion.[80] For Daston, the modern notion of algorithm can refer broadly to a step-by-step procedures and instructions used to solve problems in the mathematical way.[81] To be sure, AI algorithms fit into with this concept of algorithm. Daston flags up the hallmark of computational rules-as-algorithms.

> The key features of the thin rule are the assumptions of stability and standardization, both of conditions and users. Algorithms designed to be executed by computers are the thinnest of rules. This is not because such rules are in any sense minimalist—on the contrary, programs can be both long and complex—but because they assume complete uniformity in execution and conditions of application. Whether or not thin rules work in practice is a question of degree: whether the users and the world are more or less uniform, more or less stable.[82]

One may contest that Daston's description cannot be applied to AI because AI algorithms are dynamic. Reinforcement learning, for example, executes dynamic programming algorithms to learn behaviours and improve performance through repeatedly received reward signals. Yet, reinforcement learning can only operate 'with relatively myopic lookahead search to generate highly competent behavior' such that it is merely suitable for 'closed-world settings' rather than 'open situations that are a part of real-world environments.'[83]

Following Daston's taxonomy of rules, the rules applied to AI-powered liturgical enactment fall into the category of rules-as-algorithms. In a broader sense, rules-as-algorithms are also characteristic of the rules adopted in liturgy as scripted activity. A complete set of liturgical rules ensure that the liturgy can be properly enacted and performed in a logical order. For this reason, these liturgical rules include step-by-step instructions, which are susceptible to mathematisation. However, the liturgical script is more than rules-as-algorithms. Liturgy is intended to recapitulate the story of God's salvation and is patterned or modelled on the history of the Triune God's salvation. As such, the rules adopted in liturgy are rules-as-*models* by the virtue of the fact that they point beyond themselves to God's salvation. Looking back to the first-century churches, there were certain liturgical rules established in the church, reflecting the Christocentric and incarnational nature of liturgy (e.g., 1 Cor. 11:17–32). It can be argued that these liturgical rules of the early Church together serve as paradigms that have been emulated with the exercise of discretion throughout history.[84] Hence, the clerical enactment of liturgy involves both rules-as-algorithms and rules-as-models, and liturgical rules as models are the substance of liturgy as scripted activity. This double meaning of liturgical rules sheds light on the fact that AI ministers can partially enact liturgies to some extent but lack the capacity do so fully and authentically. That is, the two-layered meanings of liturgical rules serve as a hermeneutical lens through which to illuminate that AI ministers as the ectype of human clerics can carry out liturgical rules as *algorithms*.

Liturgical rules as models unveil a critical dissimilarity between conscious and computational liturgical enactment. With the exercise of discretion, rules-as-models can be adapted to particular circumstances and applied to particular cases. Discretion is closely associated with cognition, prudence, judgment, and other mental faculties.[85] The human mind has a unique role to play in emulating models with the exercise of discretion. The paramount aspect of the mind in relation to discretion is the consciousness of the semantic content of liturgical rules, without which liturgical rules cannot be tailored to particular circumstances. As such, a semantic capacity intrinsic to clerics distinguishes the clerical liturgical enactment from the AI-powered one.

John Searle's Chinese Room parable adds weight to the unique role of the mind in enacting rule-governed liturgies. I have demonstrated in Chapter 2 subsection 2.2.2.1 that the consciousness articulated in Searle's philosophy can be categorised as creature consciousness, which turns down the attempt made to identify human consciousness with computational or artificial consciousness. For this reason, the human mind cannot be reproduced in the silicon-based system. Searle's position is exemplified by his Chinese Room argument and marks out the uniqueness of human conscious liturgical enactment.

Searle imagines that a native English speaker locked in a room does not have any knowledge of Chinese. She is given Chinese writings, Chinese scripts, and English rules that can correlate the first two through identification and correlation of formal symbols. Moreover, she is given Chinese symbols and English instructions. With these instructions, she can correlate these Chinese symbols with Chinese writings and scripts and return new Chinese symbols in response to given Chinese symbols. In terms of output Chinese symbols, Searle judges, the native English speaker is not distinguished from native Chinese speakers. Nonetheless, she works like a symbol-processing 'computer program' and 'perform[s] computational operations on formally specified elements.'[86] For those who are outside of the room, this native English speaker is proficient in Chinese and may even be considered a native Chinese speaker. However, the computational manipulation of Chinese symbols is by no means tantamount to the understanding of Chinese.

> All that the computer has, as you have, is a formal program for manipulating uninterpreted Chinese symbols. To repeat, a computer has a syntax, but no semantics. The whole point of the parable of the Chinese room is to remind us of a fact that we knew all along. Understanding a language, or indeed, having mental states at all, involves more than just having a bunch of formal symbols. *It involves having an interpretation, or a meaning attached to those symbols.* And a digital computer… cannot have more than just formal symbols because the operation of the computer…is defined in terms of its ability to implement programs. And these programs are purely formally specifiable—that is, they have no semantic content.[87]

Searle's Chinese Room parable has sparked intense debates.[88] Objections raised to Searle's corollary are largely dependent upon the computational models of consciousness. Yet, the gravity point of Chinese Room parable is the assertion that human consciousness, understanding, interpretation, and other mental properties cannot be reproduced in the computational way precisely because it is misleading to reduce semantic content to syntactic rules and structures.

Combining Searle's Chinese Room parable with liturgical rules as both models and algorithms, it can be argued that the difference between conscious and computational enactment of liturgy lies in the fact that AI ministers are unable to comprehend the semantic content of liturgical rules as models. In this connection, Fraser Watts's observation is spot-on: 'Prayers are, after all, a fairly predictable, rule-governed "art form." [AI] could also utter prayers using a voice box. … [H]owever… they do not have the kind of inner life that would allow us to say that they really were praying. There is much more to praying than simply formulating and uttering prayers.'[89] For all that AI ministers are programmed to preach, declare God's absolution, pray for worshippers, and perform all liturgical actions, their ectypal consciousness is incapable of *understanding* the semantic content of liturgy.

The semantic content of liturgical rules underpins the meaning and goal of liturgy and, consequently, is teamed up with the intentionality of worship. For this reason, AI's incapacity to understand the semantic content of liturgical rules reveals that AI-powered liturgical enactment lacks intentions. Recent studies on the human mind also underscore the correlation between consciousness and intentionality. Some scholars suggest that the 'mental begins and ends with consciousness and intentionality,' the two together '[defining] the mental *qua* mental.'[90]

How is the correlation between consciousness and intentionality instantiated in liturgy? In his recent study on the shared intention among worshippers, Josef Quitterer extensively addresses this question, elucidating that the liturgical enactors (clerics) and liturgical participants (normal believers) share the same intentions owing to the fact that they share the same belief.[91] Following this, he registers that the shared belief in the consciousnesses of both clerics and normal believers lays the ground for the shared intentions, rather than the other way around.

> The ritual of the liturgy is not performed by an individual person and observed by passive onlookers who, through the ritual, are brought to share common beliefs.… [T]he expression 'we' in the liturgical prayer does not have the function of inducing a specific belief in the mind of the onlookers. Rather, it expresses an already existing common belief, shared by most participants. The communion of all participants presupposes common beliefs. For [these] reasons shared beliefs are not the effect but the cause of such we-expressions.[92]

Accordingly, the pronoun 'we' indicates that the worshippers's consciousness of and agreement on the shared beliefs must take place *before* they address to one another. To put it differently, the shared beliefs have already been in the worshipper's mind ahead of liturgy and are reflective of a mental state of the worshipper. In light of the nature of liturgy, the shared beliefs tie in with the recapitulation of the history of God's salvation, which is the precondition for the emergence of worshipping communities.[93]

It is this communal nature of the intentionality of liturgy that distinguishes between AI-powered and clerical enactment of liturgy and underpins the idea of AI as the ectype of humanity. The cleric's intention to enact liturgies is enclosed within the shared intentions of all worshippers. This *communal* nature has a close tie to human biological particularities (as has been elucidated above), showing an ontological distinction between AI ministers and human clerics. At the same time, the communal nature of liturgical *intentionality* foregrounds the difference between the formations of AI's and human intentions. We will return to the peculiarities of AI's intentionality in relation to moral agency in Chapter 4, subsection 4.1.3.1. Here I only touch upon how intentions are embedded within AI.

It should be conceded that AI, as well as all smart machines, are intentional in the broadest sense of the term. A real-life AI is created to perform specific tasks, such as decision-making, healthcare, and self-driving. The intentions to perform specific tasks are embedded within AI by its designers (e.g., through algorithms and data collection) and users (e.g., through the specification of functions). Machine Learning (ML) can serve as an example. Data collection and model selection are two crucial factors in ML's design. ML is trained on collected data. For this reason, the more data are collected, the better performance ML can have in the scenario of same algorithms. Data collection can be manipulated so as to meet specific demands of either individuals or communities.[94] ML models are produced by ML algorithms, which are trained on collected data and adjust their parameters to improve ML's performance on specific tasks. As such, despite the variety of models characterising the ML field as diverse, *given* tasks serve to unify the ML field.[95] Viewed in this light, both data collection and model selection show how human intentionality is built into ML, and the intentionality of ML is restricted to its performance on given tasks and cannot go beyond tasks to be addressed.

AI-powered liturgical enactment can be called intentional, but this intentionality is computational by its nature. Data concerning various Christian liturgies can be collected for training AI, and a satisfactory AI model can be produced to enact liturgies for a particular worshipping community through programming the rules-as-*algorithms* part of liturgy. In this way, worshipping communities have AI ministers trained on their own theological data. That said, AI ministers are *not conscious of* liturgy, nor is their liturgical enactment intentional. It is AI's designers and users who *intend* AI ministers to enact liturgies in a prescribed manner. In this sense, the AI minister's intentionality is the ectype of human intentional enactment of liturgy.

3.3.2.2 *Resulting in the Distinction between the Remembered and the Stored*

Given that memory is central to the mind, the contrast between human conscious and AI-powered liturgical enactment naturally gives birth to the divergence between human memory and AI's memory. The term 'memory' is widely and ambiguously used not only to describe the power of the human person's remembrance and recalling but also to measure an electronic device's capacity to store information. At this juncture, we need to recall

John Puddefoot's Sliding Definition Ploy, which has been stressed at the outset of Chapter 2, Section 2.1: some 'agree to use a word in a precise but rather unusual sense while we are on the alert, but subtly shift the meaning later to embrace everyday senses of the word not covered by their initial definition.'[96] The Sliding Definition Ploy reminds us that the different usages and concepts of memory should be kept in mind while exploring human and AI's memory.

AI cannot develop and operate without mass memory, that is, the mass storage of massive amounts of data. By way of illustration, ML's dataset has massive amounts of datapoints, which refer to specific pieces of information (e.g., images, texts, theological ideas). Each single datapoint carries many properties, and these properties are either features—which are *input variables* and can be easily computed—or labels—which are *output variables* and can only be determined and assigned by humans in correspondence with targets.[97] A hypothesis is raised as an attempt to explain or predict the relationship between features and labels and can be mathematised as a model. In order to mathematise the hypothesis, labels are assigned to training samples (to train learning algorithms), validation data (to tune the free parameters of learning algorithms trained on labelled data), and test samples (to rate learning algorithms). Free parameters or hyperparameters are selected before the training process and input into learning algorithms, and their proper values are defined in light of validation data.

Considering AI-powered liturgical enactment in light of Figure 3.1, it will come to be seen that, instead of stating that AI ministers *remember* liturgies and liturgical enactment, it is more accurate to describe AI-powered

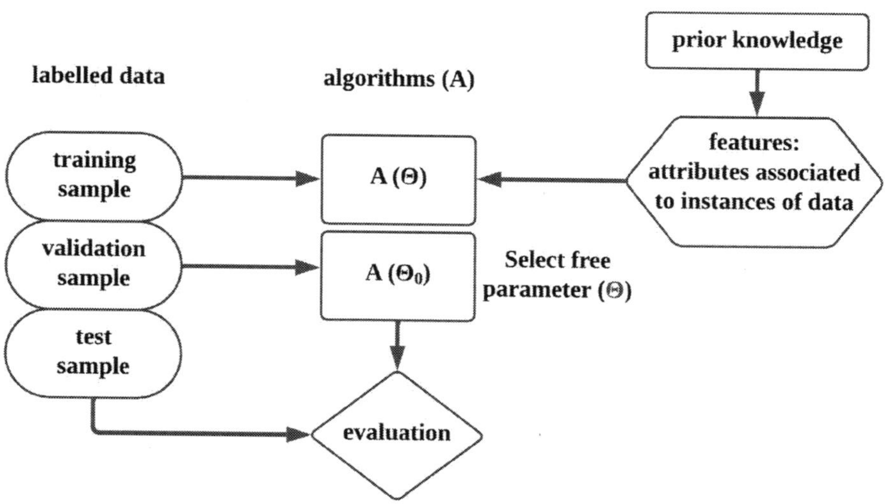

Figure 3.1 Learning stages[98]

liturgical enactment as *stored*. The features of datapoints concerning liturgical enactment can be computed and measured in accordance with human prior knowledge of Christian liturgies, such as hymns, sermons, Christian doctrines, benedictions, a theologically proper order of worship, and so forth. From the collected data of Christian liturgies are derived a number of samples or examples that are used for training, validating, and testing learning algorithms. Through data labelling, different values can be assigned to these examples in correspondence to liturgical targets. In this way, learning algorithms produce AI minister models through mathematisation of the hypothesis for explanation of the features and labelled data relevant to liturgical targets. These models can be tailored to the Christian communities of various theological traditions provided (1) that both data labelling and the prior knowledge are in light of *that* theological tradition and (2) that learning algorithms are written to exclude all data contrary to *that* theological tradition. Taking these together, AI-powered liturgical enactment is *stored* by humans into AI ministers. Even in the case of unsupervised ML—which means there is no data labelling and so no human interventions while training learning algorithms[99]—human prior knowledge of liturgy and theological-tradition-oriented algorithms together validate the corollary that AI ministers operate based both on human intentionality and on storage rather than remembrance.[100]

The clerical memory of liturgy and liturgical enactment differ from AI's storage. This distinction is manifested in the following two interrelated respects: (1) the material cause of liturgical memory, (2) the communal nature of liturgical memory.

While elucidating the contrast between the silicon-based and the biological in the above subsection 3.3.1.2, I underscored the conceptual importance of material cause for distinguishing between human and AI-powered liturgical enactment. What has been reiterated in this study is the psychosomatic unity of human beings. Hence, the idea of material cause is also of considerable importance for human memory. That is, human biological particularities are critical to both the clergy's memory and liturgical enactment.

The biological basis of memory is indisputable because, for example, some brain damage will result in the disorder of memory, showing that memory has a close tie to the nervous system.[101] Memory is related to the human biological body more broadly, and human senses and bodily action have a significant role to play in memory. The Figure 3.2 is Atkinson-Shiffrin's memory model, which was proposed more than five decades ago.[102] It presents us with a helpful lens through which to understand the cleric's memory in liturgical enactment.

This memory model illuminates that human senses are the only instrument through which information from external environments enters human memory. In the liturgical context, sensory stimuli are multiple, such as singing hymn, confession, the Eucharist, preaching, and so forth. As has been illustrated earlier, bodily worship assigns a high value to human senses, postures, gestures and other bodily actions. Through sensory registers, information

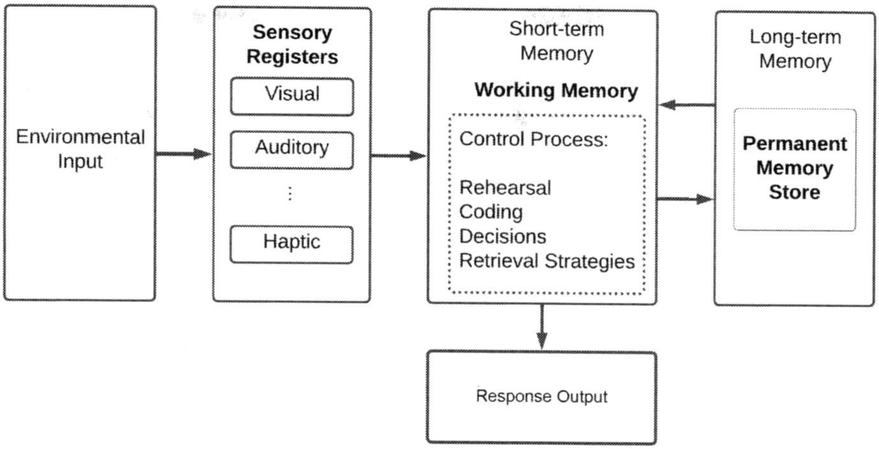

Figure 3.2 Atkinson-Shiffrin's memory model

from the liturgical context enters short-term memory, which is also named working memory.

Working memory plays an important part in the human memory system. The short-term nature of working memory means that it only has the limit capacity to store and maintain information. Yet, its functions are multiple and crucial. Thompson and Madigan observe:

> [working memory] includes retaining new information but also involves transformation and use of that information, retrieval of knowledge from long-term memory to integrate with the new information, and awareness of your surroundings. The concept of working memory is closely related to what is commonly referred to as consciousness or awareness.[103]

Reading this passage in tandem with liturgical stimuli and in light of Atkinson-Shiffrin's memory model, three points are of note here. Firstly, liturgical working memory is short-term but makes a contribution to the continuous formation of the permanent memory of liturgy. Rehearing received information will enhance short-term memory, and coding information can transform it to be easily retrievable and memorised words, phrases, and sentences.[104] Secondly, the worshipper's response to the liturgy is directly derived from working memory rather than long-term memory. That is to say, the worshipper's here-and-now response is unique and fresh on each occasion of liturgy insofar as the information from the liturgical environment is taken in worshippers through their senses on each occasion of worship. Thirdly, the worshipper's consciousness of liturgy is concomitant with the integration of the new liturgical information with the retrieved, permanent liturgical knowledge flowing from long-term memory. This new information consists of the knowledge of liturgical surroundings, including, *inter alia*, co-worshippers, buildings, space, time, and so forth.

It is worth noting that the operation of working memory has a neurophysiological basis.[105] With recent neuroimaging studies, Scott Slotnick and Daniel Schacter stress that working memory or conscious memory is tied up with neural activities in brain regions.[106] For example, the retrieval of knowledge from long-term memory is most prominently associated with neural activity in the prefrontal cortex. Having stressed above that neural activities belong into the bodily significance of worship, I will focus attention on the function of mirror neurons in relation to liturgical memory.

Mirror neurons are a subset of visuomotor neurons, and their two main functions are understanding and imitating the actions of others.[107] With the mirror mechanism, the sensory representations about observed actions are transformed into motor representations concerning actions that the observer would perform if she plans to do so. This is so because the brain receives two pieces of information when one is observing an action performed by others: what the action is, and why the action is performed. The former means the observer *understands* the action, and the latter means the observer knows the *intention* of the action.[108] With these two pieces of information, the mirror mechanism enables humans to understand and imitate actions of others.

David Hogue correlates mirror neurons studies and liturgy, showing the communal nature of worship and the importance of bodily worship. Rather than simply identifying the brain as a platform for the operations of the minds, he argues, mirror neurons studies shed light on the facts both that 'the brain [is] an organ of connection with others, of relationship [and that our] brains are particularly designed to make sense of other human beings, their thoughts and feelings, and to respond in ways that communicate understanding and compassion.'[109] Here the emphasis falls on the role that mirror neurons play in human participation into liturgy. The following lengthy claim made by Hogue is worth quoting to elucidate how mirror neurons operate in liturgy.

> [S]imply hearing the sounds of an action (like the tearing of a piece of paper) will stimulate motor neurons in the hands that would be necessary to perform the same action. Even more relevant to our practices of preaching and worship, they persuasively argue that *hearing descriptions of actions stimulates not only the neurons re-sponsible for speech (imitating the speech of the speaker), but even the motor neurons that would produce the actions being described*. So when we are listening to a story or a sermon, *our brains are mimicking not only the actions of the preacher or presider, but the actions of the characters in the stories being told*. Either seeing or hearing an event prompts the brain to imagine its own body performing a particular action. We rehearse for ourselves what we see and hear others doing.[110]

In Hogue's position, an individual's worship took and is taking shape within the worshipping community through operation of mirror neurons. With Atkinson-Shiffrin's memory model, we arrive at the corollary that through human senses the mirror mechanism takes the information from liturgical environments into working memory, which, in turn, strengthens long-term liturgical memory with rehearsal, imitation, imagination, and so forth.

I have hitherto demonstrated the biological particularities of the worshipper's memory of liturgy with particular attention to the functioning of mirror neurons. Needless to say, these particularities feature in the clerical enactment of liturgy in that, as noted in the subsection 3.2.3, the clergy become part of the worshipping community before enacting liturgies. In this light, the clerical enactment of liturgy is bound up with and shaped in the clergy's liturgical memory. As an illustration, the clergy's memory of being addressed in worship must underlie their action of addressing to worshippers, and the mirror mechanism operates to enable them to imitate and perform afresh God's address to worshippers delivered by other clerics. Furthermore, worshippers's real-time responses to liturgy have an impact on the clergy's liturgical memory. The mirror mechanism in the cleric's brain is a biological foundation for the cleric's understanding of the worship of the community which she belongs into. David Hogue rightly notes that by virtue of mirror neurons studies, '[l]iturgical leadership requires pastoral sensitivity to the profound ways in which culture, class, and gender both enrich and constrain possibilities for vital worship.'[111] This sensitivity is partly predicated upon the mirror neurons in that the mirror mechanism enables the clerics to know the intention of worshippers's liturgical action. As such, the clerical enactment of liturgy is indexed to the faith communities of different cultures, circumstances and traditions, and the cleric's actions of liturgical enactment are fresh and unique on each occasion of worship and for each worshipping community.

The human memory system operates in a different way than AI's storage system. Admittedly, there are similarities between the AI's storage and human remembrance of liturgy, say, information processing. However, clerical and AI-powered enactment of liturgy differ in that the clergy's biological particularties and neural activities lay a foundation for their memory of liturgy, which has been shaped before liturgical enactment. Rather than being stored as what happens to AI, the liturgy patterned on the history of God's salvation is remembered and enacted by human clergy.

3.3.3 Summary

Does AI enhance or weaken liturgical enactment? The answer to this question is not all-or-none. From the vantage points of both bodily and conscious enactment and considering the findings of the first two chapters, AI-powered liturgical enactment can be regarded as the ectype of the clerical enactment of liturgy. The bodily nature of liturgy cannot rule out the possibility to apply AI to liturgical enactment, nor does the clergy's conscious enactment of liturgy put an end to AI-powered liturgical enactment. Considering AI as the *imago hominis*, AI-powered liturgical enactment can be justified in the sense of ectypal action, which discloses how AI participates into the stories of God and humans.

3.4 AI-Powered Liturgical Enactment to Extend Worship

Discarding the pursuit of an all-or-none answer to the question of AI-powered liturgical enactment, the idea of AI as the ectype of humanity sets a scene for an appreciative yet critical reception of AI ministers within liturgical communities. That is, AI-powered liturgical enactment is, on no account, tantamount to the clerical enactment of liturgy but is capable of *extending* worship widely.

Liturgical extension implies that the Christian liturgy is restricted within certain spatiotemporal confines. As noted in the subsection 3.2.2.1, human senses, bodily gestures and postures used in liturgy are reflective of the spatiotemporality of liturgy. Theologically speaking, this spatiotemporality is rooted in the fact that the liturgy is patterned on the history of salvation, the centre of which is the incarnation. Viewed in the light of the incarnation, spatiotemporality is more a upside than a downside of Christian liturgy. In this regard, Thomas Torrance's treatment of the incarnation in relation to space and time offers a cogent contention:

> in Jesus Christ God Himself in His own Being has come into our world and is actively present as *personal Agent within our physical and historical existence*. As both God of God and Man of man Jesus Christ is the actual Mediator between God and man and man and God in all things, even in regard to space-time relations. He constitutes in Himself the rational and personal Medium in whom God meets man in *his creaturely reality* and brings man *without having to leave his creaturely reality* into communion with Himself.[112]

In Torrance's position, the incarnation bestows a special value upon human physical existence and underscores the redemptive significance of human creaturely reality, which is restricted within spatiotemporal confines. Human creaturely reality also participates in communion with God. Following this, he contends that the doctrines of incarnation and creation should be teamed up for speaking of God who is present to us within space and time as 'the rational medium.'[113] By virtue of the incarnation, human creaturely reality carries both spatiotemporal and relational connotations in terms of their communion with God.

Torrance's incarnational account of space–time relations sheds much light on liturgical extension as a *positive* consequence of liturgical spatiotemporality. Liturgy is one of the most vivid examples of human communion with God, in which their creaturely reality—including both their bodies and minds—deeply involves. Patterned on the history of salvation, the liturgy is integrated with the incarnational validation of spatiotemporal creaturely reality.[114] For this reason, human creaturely reality in a particular space and time is a *sine qua non* for liturgy. It is this spatiotemporal nature of liturgy that undergirds communal worship. That is, a group of worshippers gather together in a space and time with intention to worship God, and both space and time are favourable ingredients of liturgy.[115]

In the light of this positive appraisal of both creaturely reality in worship and liturgical spatiotemporality, liturgical extension should not be construed as a way in which the church seeks to overcome the reality of human creaturely limitations. Rather, liturgical extension is an *invitation* that is sent to those who are unable to participate into a liturgy enacted in a particular space and time. Coupled with the idea of AI as the ectype of humanity, this affirmative description of liturgical extension furnishes us with a hermeneutical lens through which I propose three principles for AI-powered liturgical enactment.

First, AI-powered liturgical enactment is *authentic* and yet *ectypal and subaltern* due to its lack of holistic humanity. Although the holistic reading of the *imago Dei* foregrounds the importance of both the body and the mind for liturgical enactment, it is nevertheless a misleading and harsh claim that AI-powered liturgical enactment should be nullified altogether. My analysis of the contrasts between clerical and AI-powered liturgical enactment suffices to show that the history of God's salvation is somehow recapitulated in the liturgy enacted by AI ministers. To this extent, AI-powered liturgical enactment is real and authentic without any doubt. This corollary is partly resonant with Katherine Schmidt's recent study on Christian sacraments in the virtual world.[116] Considering the dialectic of presence and absence as the essence of virtuality, Schmidt asserts that the internet and digital technology can serve to mediate Christian sacraments in virtual space insofar as God's presence and absence lie at the centre of Christian sacramental and ecclesial life.[117] In other words, human presence yet absence in AI-powered virtual liturgy can be justified by the virtue of the recapitulation of the salvation made by the present yet absent God. While concurring with Schmidt's justification of the authenticity of virtual sacramental life, I submit a more nuanced construal of virtuality both in the theological and technological sense. Technological virtuality draws our attention not only to human embodiment or materiality (which Schmidt has discussed) but also to holistic humanity, biological particularities in particular, in relation to liturgical enactment.

AI-powered liturgical enactment should, therefore, be qualified in an ectypal sense. It is ectypal in that AI mirrors human clerics and extends liturgies across space and time. In this way, liturgical spatiotemporality is not eliminated but rather is displayed in variegated fashions. This ectypal implication leads to the subaltern nature of AI-powered liturgical enactment. Although the idea of AI as the ectype of humanity brings forth the close connection between clerics and AI ministers, the ectypal implication presupposes the ontological distinction between the archetype and the ectype. As such, however valuable and powerful is AI in mediating liturgies, AI-powered liturgical enactment as *subaltern enactive action* is overshadowed when compared to the human clerical enactment of liturgy. That is to say, AI-powered liturgical enactment should give way to the human clerical enactment of liturgy in the event that the latter is viable.

Having this ectypal and subaltern implication in mind, we can sketch a spectrum outside which AI-powered liturgical enactment is ill-grounded. There are multifaceted applications of AI to liturgy, as exemplified in the above Section 3.1. The examples of AI ministers at St. Paul's Church in Nuremberg, Germany, is at odds with the ectypal and subaltern nature of AI-powered liturgical enactment. It lacks the reason to enact the liturgy by AI ministers precisely because human creaturely reality and liturgial spatiotemporality can be better expressed through the viable clerical enactment of liturgy. It can be argued that such a subaltern enactive action as a substitute for human enactment underlies some worshippers's dissatisfaction with the AI-driven liturgy at St. Paul's Church. After all, it is the human minister as the archetype, rather than AI ministers as the ectype, who participates in liturgy both bodily and mentally before enacting the liturgy.

Concomitant with the first principle is the second one: AI-powered liturgical enactment is used to extend the invitation for the liturgy broadly by virtue of the fact that AI as the ectype of humanity is embedded with the capacity to augment liturgies. The idea of AI as the ectype of humanity offers a theological apparatus through which to justify the AI-powered augmentation of liturgy. Following the ontological connection between human clerics and AI ministers, AI-powered enactive action is considered to be the ectype of human clerical enactment and capable of carrying theological significance of liturgy. Behind the ectypal enactive action is the silicon-based system that stores liturgical information and executes those liturgical rules as algorithms. In this manner, AI extends liturgies beyond human biological limitations and mediates them broadly to others through virtual reality (VR). An example of this type is VR Church, which is a virtual community and enacts liturgies in the metaverse.[118] VR Church has human pastors, and each pastor has his or her avatar in the virtual world. Although VR Church does not have AI ministers to enact liturgy as what happened in St. Paul's Church, its liturgical enactment can still be considered AI-powered in a broad sense because of the metaverse's reliance upon AI technology (e.g., creation of virtual world). Some parts of liturgy (e.g., the sacraments) can only be enacted in AI-assisted way.

That said, liturgical extension beyond biological limitations should be construed in conjunction with the value of both human creaturely reality and liturgical spatiotemporality. As such, liturgical extension under the auspice of AI-powered augmentation can be viewed as the liturgical invitation sent widely to those who are unable to participate in liturgy *due to their creaturely reality confined within a particular space and time*. This imagination of AI-powered liturgical enactment can be proved effective in the time of the COVID-19 pandemic. During the pandemic, digital technologies have been utilised to uphold churchly communion and extend church ministries.[119] Likewise, AI can enhance ecclesial life and liturgical enactment during times of crisis. In short, AI-powered liturgical enactment is, in essence, a liturgical invitation by which worshippers's creaturely reality touches the liturgy in an extended and augmented way.

The first two principles together give rise to the third one: AI-powered liturgical enactment aims at orienting individuals to participate into communal liturgy in a particular space and time. Given (1) that AI-powered liturgical enactment is ectypal and subaltern, and (2) that AI-powered liturgical extension is an invitation, AI-powered liturgical enactment seeks solely to consolidate the communal and spatiotemporal liturgy. In any case, the ectypal enactive action cannot fully unveil liturgical meaningfulness, which is intertwined with holistic humanity.

AI-powered liturgical enactment undeniably augments liturgies and extends the liturgical invitation to participate. However, it also has the potential to breed liturgical individualism, which means that individuals perform liturgies on their own. For example, an individual can participate into liturgies in the metaverse with a VR headset. Such an easy access to liturgies may induce worshippers to turn away from a spatiotemporal worshipping community while preoccupying themselves with personal liturgical preferences or customised liturgies, which is at variance with the communal nature of liturgy as expressed through holistic humanity (see Section 3.2). Furthermore, AI-powered liturgical enactment carries a social implication insofar as the *imago hominis* mirrors the collective significance of the *imago Dei* (see subsection 3.3.1.1). Hence, this liturgical individualism induces AI-powered liturgy to operate off the track. At this point, Harris Wiseman's detailed discussion over liturgical meaningfulness underwrites the third principle over against this individualistic abuse of AI-powered liturgical enactment:

> Meaningfulness arises between the persons engaged in the ritual, one another, and God. The meaningfulness of that connectedness is not in the words or actions of the ritual, but in their being performed together at that moment. In that corporate regard, at least, the words and gestures are not codes at all. Rather, they are the means, the shared structure, the order, by which the meaningfulness of performing a ritual together can arise. In corporate ritual, worshipping others are known directly as copresences (rather than just as objects of theoretical knowledge). In addition to having ideas about the world, and the others one knows, one also knows them directly, by being with them and alongside them.[120]

Accordingly, the meaningfulness of AI-powered liturgical enactment lies in its valuable orientation of individuals towards communal worship.

These three principles together can steer the course of future liturgical extension augmented by AI in order not to downplay the importance of both the human biological body and the mind for enacting liturgies. In this way, AI-powered liturgical enactment is tailored to the recapitulation of the history of salvation and, consequently, brings to the foreground the Christological nature of liturgy, that is, the incarnation.

3.5 Conclusion

It has been demonstrated thus far how Christian liturgical enactment serves as a litmus test for the findings of the first two chapters: the holistic reading of the *imago Dei* gives birth to the psychosomatic meaning of being human and the idea of AI as the *imago hominis* (the ectype of humanity). The ontological distinction yet connection between the *imago Dei* and the *imago hominis* are exhibited in the contrast between human clerical and AI-powered enactment of liturgy.

The two pairs of the contrasts examined in this chapter highlight the uniqueness of human clerics in liturgical enactment vis-à-vis AI. Derived from the two pairs are four sets of divergences between clerical and AI-powered enactment of liturgy: (1) the real versus the virtual, (2) the biological versus the silicon-based, (3) the conscious versus the computational, and (4) memory versus storage. As the Figure 3.3 shows, the ontological implication of archetype–ectype lies at the centre of the four sets of divergences.

The four sets of divergences display a multi-layered relationship between human clerical and AI-powered liturgical enactment. The solid line between the real and the virtual means that they are connected rather than isolated. The distinction between virtuality and reality cannot be simply employed to rule out the possibility of applying AI to liturgy. Rather, AI can augment liturgies and extend them through the virtual to reach the real. The dotted line between the silicon-based and the biological indicates the subaltern nature of AI-powered enactment of liturgy. After all, the silicon-based system cannot replace the human body by reason of the latter's biological particularities. The dashed lines between the rest two sets of divergences connote that AI-powered liturgical enactment is temporary and orients worshippers to in-person communal and spatiotemporal liturgies.

The four sets of divergences between clerical and AI-powered liturgical enactment showcase that the liturgy centred in the incarnation can neither be reduced to information or data nor can it be simply enacted through the mathematical operations of algorithms. Rather than correlating AI-powered liturgical enactment with AGI or ASI, we need to probe how AI can be applied to liturgical enactment with intention to embody the theological implications of liturgy. As the ectype of humanity, AI reminds us of the importance of psychosomatic humanity in liturgical enactment. At the same time, AI-powered

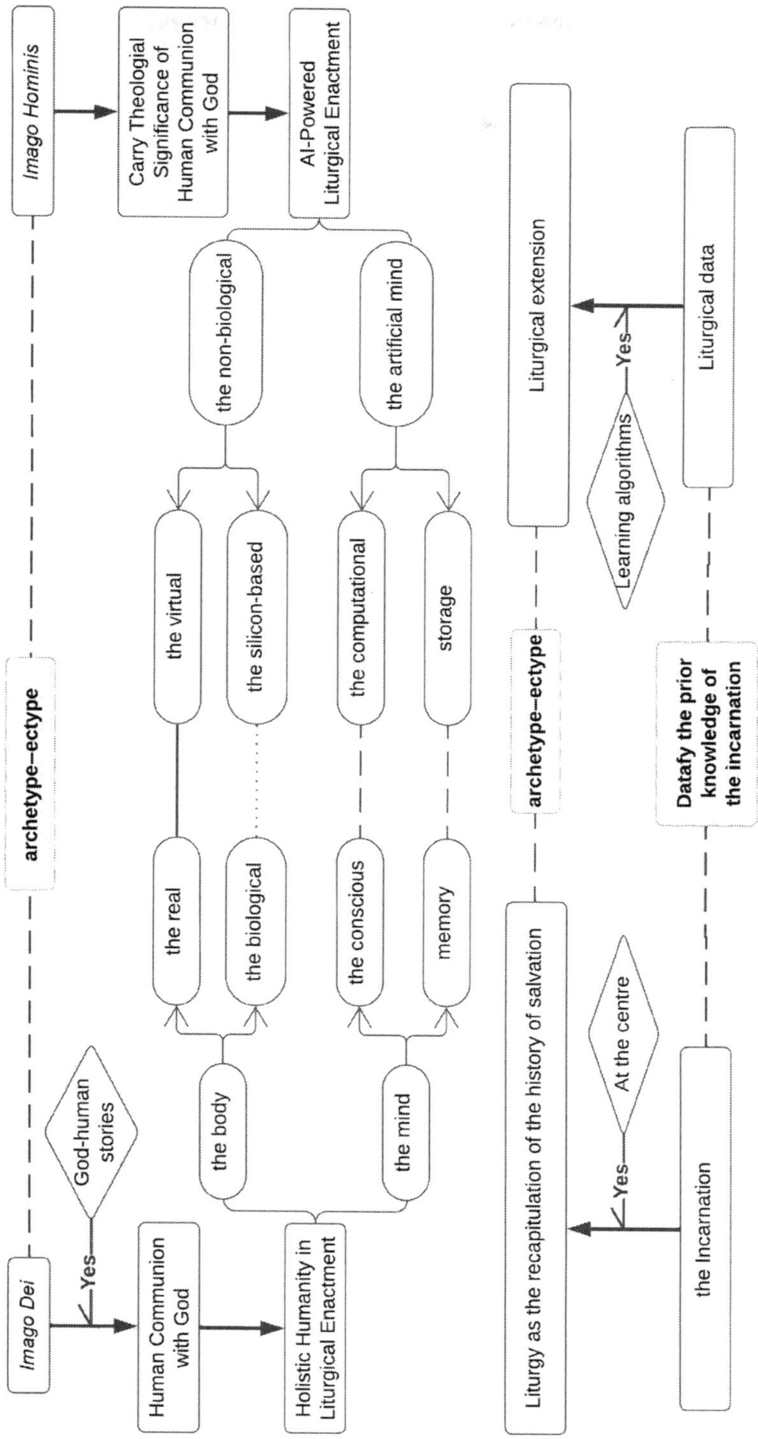

Figure 3.3 Liturgical enactment: Clerics vs AI

liturgical extension indicates that the meaningfulness of AI-powered liturgical enactment lies in AI's augmentation of the recapitulation of the history of God's salvation, that is, the redemptive stories of God and humans. To this extent, AI as the *imago hominis* participates in the religious performances made by the *imago Dei* towards God.

Notes

1 Geoffrey Wainwright, "Christian Worship: Scriptural Basis and Theological Frame," in *The Oxford History of Christian Worship*, ed. Geoffrey Wainwright and Karen B. Westerfield Tucker (Oxford: Oxford University Press, 2005), 9–19.
2 Frank C. Senn, *Christian Liturgy: Catholic and Evangelical* (Minneapolis: Fortress, 1997), xiv.
3 Susan White, *Christian Worship and Technological Change* (Nashville: Abingdon, 1994).
4 See, e.g., Doug Gay and Ron Rienstra, "Veering Off the Via Media: Emerging Church, Alternative Worship, and New Media Technologies in the United States and United Kingdom," *Liturgy* 23, no. 3 (2008): 39–47; Kimberly H. Belcher, Kevin G. Grove, CSC, and Sonja K. Pilz, "Recording as the Re-Membering Work of the People: A Catholic-Jewish Dialogue on the Body and Liturgical Memory," *Studia Liturgica* 51, no. 2 (2021): 122–142.
5 See, e.g., Young, "Reverend Robot," 479–500; Sarah Kathleen Johnson, "Online Communion, Christian Community, and Receptive Ecumenism: A Holy Week Ethnography during COVID-19," *Studia Liturgica* 50, no. 2 (2020): 188–210.
6 Teresa Berger, *@ Worship: Liturgical Practices in Digital Worlds* (London: Routledge, 2018).
7 Philip Hefner, "The Church," in *Christian Dogmatics, Volume 1 & 2*, ed. Carl E. Braaten and Robert W. Jenson (Philadelphia: Fortress, 2011), 2:231.
8 Friedrich Schleiermacher, *On Religion: Speeches to Its Cultured Despisers*, ed. and trans. Richard Crouter, 2nd ed. (Cambridge: Cambridge University Press, 1996), 93.
9 Adam Macinnis, "An AI Aims to be First Christian Celebrity of the Metaverse," *Christianity Today*, 14 February 2022, https://www.christianitytoday.com/ct/2022/march/ai-artificial-intelligence-music-metaverse-gospel-singer-jc.html.
10 Robin A. Parry, *Worshipping Trinity: Coming Back to the Heart of Worship*, 2nd ed. (Eugene: Cascade Books, 2012), 98–117.
11 Kirstern Grieshaber, "Can A Chatbot Preach A Good Sermon? Hundreds Attend Church Service Generated By Chatgpt To Find Out," *AP News* 10 June 2023, https://apnews.com/article/germany-church-protestants-chatgpt-ai-sermon-651f21c24cfb47e3122e987a7263d348.
12 J. D. Crichton, "A Theology of Worship," in *The Study of Liturgy*, ed. Cheslyn Jones, Geoffrey Wainwright, and Edward Yarnold, SJ (London: SPCK, 1978), 18–19.
13 Jean-Jacques von Allmen, *Worship: Its Theology and Practice* (New York: Oxford University Press, 1965), 33. The idea of the liturgy as the recapitulation of the history of salvation also features in other works on Christian liturgies, see, for example, Edward J. Kilmartin, *Systematic Theology of Liturgy*, Christian Liturgy: Theology and Practice, Volume I (Kansas: Sheed & Ward, 1988), 53–90; James Farwell, *This Is the Night: Suffering, Salvation, and the Liturgies of Holy Week* (London: T&T Clark, 2005), 93–134.
14 Ian A. McFarland, *The Word Made Flesh: A Theology of the Incarnation* (Louisville: Westminster John Knox, 2019), 3.

15 An example of this type is G. W. F. Hegel, "The Positivity of the Christian Religion," in *On Christianity: Early Theological Writings* trans. T. M. Knox (New York: Harper & Brothers, 1961), 67–85.
16 Ola Sigurdson, *Heavenly Bodies: Incarnation, the Gaze, and Embodiment in Christian Theology*, trans. Carl Olsen (Grand Rapids, MI: Eerdmans, 2016), 47.
17 Ibid., 20, 147; emphasis added.
18 Abraham Kuyper, *Our Worship*, trans. Harry Boonstra et al., ed. John D. Witvliet and Harry Boonstra (Grand Rapids: Eerdmans, 2009), 142. He also correlated worship and a holistic view of human being: 'A person, consisting of soul and body, who fears God, has a need to express his religious feelings in two ways—through words and actions. The first one concerns his spirit, and the second one his senses' (216). A similar viewpoint can also be found in Nicholas Wolterstorff, *The God We Worship: An Exploration of Liturgical Theology* (Grand Rapids, MI: Eerdmans, 2015), 24.
19 Kilmartin, *Systematic Theology of Liturgy*, 19.
20 Ibid., 20–21.
21 Ibid., 61.
22 Ibid., 26.
23 Ibid.
24 Frank C. Senn, *Embodied Liturgy: Lessons in Christian Ritual* (Minneapolis: Fortress, 2016), 1.
25 Ibid., 15.
26 Ibid., xii. He also argues elsewhere: 'Postures are also symbolic uses of the body that express the reality of what the worshipers are experiencing.' Frank C. Senn, *Introduction to Christian Liturgy* (Minneapolis: Fortress, 2012), 192.
27 Senn, *Embodied Liturgy*, 72; emphasis added. A similar claim is made by David Torevell: 'We "learn" to share in the divine life of Christ primarily by the engagement of our bodily senses, not by increasing attention being given to the stimulation of our minds. The power of ritual is achieved through the symbolic use of the body. The nature of liturgical symbol means that in order for us to be participants, the whole human being must be engaged through the senses.' David Torevell, *Losing the Sacred: Ritual, Modernity, and Liturgical Reform* (London: T&T Clark, 2004), 194–195.
28 Carl E. Braaten and LaVonne Braaten, *The Living Temple: A Practical Theology of the Body and the Foods of the Earth* (Eugene, Oregon: Wipf & Stock, 1976), 6.
29 Fraser Watts, "Embodied Cognition and Religion," *Zygon: Journal of Religion & Science* 48, no. 3 (2013): 747.
30 Fraser Watts, *A Plea for Embodied Spirituality: The Role of the Body in Religion* (London: SCM, 2021), 95–107.
31 John A. Teske, "From Embodied to Extended Cognition," *Zygon: Journal of Religion & Science* 48, no. 3 (2013): 764–767.
32 Robert C. Fuller and Derek E. Montgomery, "Body Posture and Religious Attitudes," *Archive for the Psychology of Religion* 37 (2015): 227–239.
33 Richard F. Thompson and Stephen A. Madigan, *Memory: The Key to Consciousness* (Princeton: Princeton University Press, 2005), v, 1.
34 Peter Atkins, *Memory and Liturgy: The Place of Memory in the Composition an Practice of Liturgy* (Aldershot: Ashgate, 2004), 25; emphasis added. Bruce Morrill holds a similar yet more vigorous argument that memory is central to the practice of the Christian faith. Bruce T. Morrill, S. J., *Anamnesis as Dangerous Memory: Political and Liturgical Theology in Dialogue* (Collegeville, MN: Liturgical Press, 2000), 139–146.
35 Atkins, *Memory and Liturgy*, 65.
36 Ibid., 11. Frank Senn also argues in a similar way: 'Where there is memory, there is the possibility of retrieval. This retrieval may be facilitated through the

performance of ritual actions, since the sacramental signs connect us with life and the world, and the sacramental celebration connect us with the memory of the past … and hope for the future ….' Senn, *Christian Liturgy*, 7.

37 Julie Gittoes, *Anamnesis and the Eucharist: Contemporary Anglican Approaches* (Aldershot: Ashgate, 2008), 78. Marva Dawn also suggests that the liturgy helps the church to relearn the language of eschatology; Marva J. Dawn, *A Royal "Waste" of Time: The Splendor of Worshiping God and Being Church for the World* (Grand Rapids, MI: Eerdmans, 1999), 362–364.

38 Atkins, *Memory and Liturgy*, xiii.

39 This also echoes Gregory of Nazianzus's view of worship, summarised by N. V. Harrison as follows: 'Anamnesis means representation of God's saving works so that the worshipers can participate in these events as present realities and thereby receive the eschatological salvation, new life and sanctification divinely accomplished through them. Anamnesis thus unites past, present and future in a single present event of worship.' N. V. Harrison, "Introduction," in *Gregory of Nazianzus, Festal Orations* (Crestwood: St. Vladimir's Seminary Press, 2008), 24.

40 Joshua Cockayne and Gideon Salter, "Feasts of Memory: Collective Remembering, Liturgical Time Travel and the Actualisation of the Past," *Modern Theology* 37, no. 2 (2021): 295.

41 Atkins, *Memory and Liturgy*, 82.

42 Wolterstorff, *The God We Worship*, 11.

43 Joseph Ratzinger, *The Spirit of the Liturgy*, trans. John Saward (San Francisco: Ignatius Press, 2000), 80.

44 Wolterstorff, *The God We Worship*, 66.

45 Ibid., 61.

46 von Allmen, *Worship*, 133.

47 Ibid., 143.

48 Ibid., 138.

49 It is interesting that Fran Senn discusses the connection between the clergy's function and what they wear in liturgy; Senn, *Introduction to Christian Liturgy*, 186–188.

50 "We Brought Church to the Metaverse," Life.Church, https://www.life.church/metaverse/.

51 Young, "Virtual Pastor," 6–22; Young, "Reverend Robot," 479–500.

52 Berger, *@ Worship*, 8.

53 Ibid., 16.

54 Ibid.

55 Ibid., 18.

56 Ibid., 19.

57 Philip Brey, "The Social Ontology of Virtual Environments," *American Journal of Economics and Sociology* 62, no. 1 (2003): 276–277.

58 Luciano Floridi, *The Fourth Revolution: How the Infosphere Is Reshaping Human Reality* (Oxford: Oxford University Press, 2014), 218.

59 Massimo Durante, "Technology and the Ontology of the Virtual," in *Oxford Handbook of Philosophy of Technology*, ed. Shannon Vallor (Oxford: Oxford University Press, 2022), 319.

60 Ibid., 336.

61 William A. Dyrness, *A Primer on Christian Worship: Where We've Been, Where We Are, Where We Can Go* (Grand Rapids: Eerdmans, 2009), 114–115.

62 The sociality of technology is widely recognised; see, for example, Oliver O'Donovan, *Begotten or Made?* (Oxford: Clarendon, 1984), 6; Noreen Herzfeld, *Technology and Religion: Remaining Human in a Co-created World* (West Conshohocken: Templeton Press, 2009), 8; Jacqui Stewart, "Technology and Christianity," in *God, Humanity and the Cosmos: A Textbook*

in Science and Religion, ed. Christopher Southgate (London: T&T Clark, 2011), 372.
63 Yuval Noah Harari, *Homo Deus: A Brief History of Tomorrow* (London: Vintage, 2017), 83.
64 Ibid., 84–87.
65 For example, ibid., 117, 158, 387.
66 Antonio R. Damasio, *The Strange Order Of Things: Life, Feeling and the Making of Cultures* (London: Robinson, 2021), 200–201.
67 Ibid., 201; emphasis added.
68 Ibid., 65–68.
69 Simon Balle and Ulrik Nissen, "Responsive Bodies: Robots, AI, and the Question of Human Distinctiveness," *Zygon: Journal of Religion & Science* 58, no. 2 (2023): 363.
70 Ibid., 369–369. Philosophically speaking, responsiveness belongs to the existential feature of the human condition; see Thomas Schwarz Wentzer, "Approaching Philosophical Anthropology: Human, the Responsive Being," in *Finite but Unbounded: New Approaches in Philosophical Anthropology*, ed. Kevin M. Cahill, Martin Gustafsson, and Thomas Schwarz Wentzer (Berlin: De Gruyter, 2017), 25–45.
71 Balle and Nissen, "Responsive Bodies," 367–368.
72 Wolterstorff, *The God We Worship*, 5–6.
73 Ibid., 7.
74 Nicholas Wolterstorff, *Acting Liturgically: Philosophical Reflection on Religious Practice* (Oxford: Oxford University Press, 2018), 14.
75 Wolterstorff, *The God We Worship*, 15; emphasis original.
76 Wolterstorff, *Acting Liturgically*, 23; also see Wolterstorff, *The God We Worship*, 7; Nicholas Wolterstorff, "Knowing God Liturgically," *Journal of Analytic Theology* 4 (2016): 5–7.
77 Wolterstorff, *The God We Worship*, 8; Wolterstorff, *Acting Liturgically*, 23.
78 Lorraine Daston, *Rules: A Short History of What We Live By* (Princeton: Princeton University Press, 2022), 151–237.
79 Ibid., 23–81.
80 Ibid., 82–150.
81 Ibid., 86, 92, 94, 103–104, 126.
82 Ibid., 79–80.
83 Russell, *Human Compatible*, 55; Landgrebe and Smith, *Why Machines Will Never Rule the World*, 174.
84 Further on the origins and features of Christian worship in the early Church, see Larry W. Hurtado, *At the Origins of Christian Worship: The Context and Character of Earliest Christian Devotion* (Grand Rapids: Eerdmans, 1999).
85 Daston presents us with a helpful survey of the idea of discretion; Daston, *Rules*, 36–40.
86 Searle, "Minds, Brains, and Programs," 417–418.
87 Searle, *Minds, Brains and Science*, 33; emphasis added.
88 For example, John Preston and Mark Bishop, eds., *Views into the Chinese Room: New Essays on Searle and Artificial Intelligence* (Oxford: Clarendon, 2002); McDermott, "Artificial Intelligence and Consciousness," 141–144.
89 Fraser Watts, "Artificial Intelligence," in *God, Science and Humility: Ten Scientists Consider humility Theology*, ed. Robert L. Herrmann (Philadelphia: Templeton Foundation Press, 2000), 284.
90 George Graham, Terence Horgan, and John Tienson, "Consciousness and Intentionality," in *The Blackwell Companion to Consciousness*, ed. Susan Schneider and Max Velmans (Chichester: Wiley Blackwell, 2017), 519. It is beyond scope to discuss this correlation here; a typical work on consciousness and intentionality is John R. Searle, *Intentionality: An Essay in the Philosophy of Mind* (Cambridge: Cambridge University Press, 1983).

91 Josef Quitterer, "Liturgical Rituals as Shared Intentional Practices," Studia Liturgica 48, no. 1-2 (2018): 56. As such, Quitterer submits that there is no asymmetric relationship between clerics and normal believers during liturgy.
92 Ibid., 58.
93 This reinforces Quitterer's claim:

> liturgical celebrations are the expression of a close cooperation of more or less active participants of a group on the basis of shared beliefs. These beliefs are not merely the aggregate of the mental states of the single group-members. They are beliefs which occur within a group that has already (explicitly or implicitly) agreed to share specific beliefs, goals and intentions.
>
> Ibid., 61.

94 Kate Crawford spends much space discussing that data collection for training AI 'is a social and political intervention' rather than 'a purely technical one;' Kate Crawford, Atlas of AI: Power, Politics, and the Planetary Costs of Artificial Intelligence (New Haven: Yale University Press, 2021), 89–121.
95 Peter Flach, Machine Learning: The Art and Science of Algorithms that Make Sense of Data (Cambridge: Cambridge University Press, 2012), 13.
96 Puddefoot, God and the Mind Machine, 57.
97 Further on features and labels, see Alexander Jung, Machine Learning: The Basics (Singapore: Springer, 2022), 19–30.
98 This figure is taken from Mehryar Mohri, Afshin Rostamizadeh, and Ameet Talwalkar, *Foundations of Machine Learning*, 2nd ed. (Cambridge, Massachusetts: MIT Press, 2018), 5.
99 According to Alexander Jung, there are two large families of unsupervised machine learning methods. First, the clustering learning method means to classify datapoints according to their similarities into subsets. Second, the feature learning method means to datapoints can be processed efficiently in light of their numeric features; Jung, *Machine Learning*, 153–188.
100 It has been widely recognised that algorithms are not neutral but very likely to be embedded with values and biases. That is to say, learning algorithms can be designed in the direction of a particular theological tradition.
101 For example, amnesia is partly related to brain damage; Alan D. Baddeley, *Essentials of Human Memory* (East Sussex: Psychology Press, 1999), 193–207.
102 Richard C. Atkinson and Richard M. Shiffrin, "The Control of Short-Term Memory," *Scientific American* 225, no. 2 (1971): 82. Atkinson and Shiffrin's model of memory was first raised in Richard C. Atkinson and Richard M. Shiffrin, "Human Memory: A Proposed System and Its Control Processes," in *The Psychology of Learning and Motivation*, ed. K. Spence and J. Spence (New York: Academic Press, 1968), 89–105. Admittedly, this model has been criticised from various perspectives, but its basics 'have stood the test of time'; Thompson and Madigan, *Memory*, 42. They offer a critical assessment of this model, see 25–49.
103 Thompson and Madigan, *Memory*, 12.
104 Atkinson and Shiffrin, "The Control of Short-Term Memory," 83.
105 Atkinson and Shiffrin are aware that working memory involves the functioning of the brain, but they do not expand on this point; ibid.
106 Scott D. Slotnick and Daniel L. Schachter, "The Cognitive Neuroscience of Memory and Consciousness," in *The Cambridge Handbook of Consciousness*, ed. Philip David Zelazo, Morris Moscovitch, and Evan Thompson (Cambridge: Cambridge University Press, 2007), 809–827. Elsewhere Slotnick extensively investigates the association between different types of memory and brain regions. Scott D. Slotnick, *Cognitive Neuroscience of Memory* (Cambridge: Cambridge University Press, 2017).
107 Giacomo Rizzolatti and Laila Craighero, "The Mirror-Neuron System," *Annual Review of Neuroscience* 27 (2004): 169–192.

108 Giacomo Rizzolatti, "The Mirror Neuron System and Its Function in Humans," *Anatomy and Embryology* 210, no. 5–6 (2005): 420.
109 David A. Hogue, "Sensing the Other in Worship: Mirror Neurons and the Empathizing Brain," *Liturgy* 21, no. 3 (2006): 32.
110 Ibid., 36; emphasis added.
111 Ibid., 36.
112 Thomas F. Torrance, *Space, Time and Incarnation* (Edinburgh: T&T Clark, 1997), 52; emphasis added. Torrance's relational understanding of space and time based on the doctrine of the incarnation is critically examined by Antje Jackelén, who works out a different relational model of space and time to get around dualism; Antje Jackelén, *Time and Eternity: The Question of Time in Church, Science, and Theology* (Philadelphia: Templeton Foundation Press, 2005), 186–190. For the purpose of this chapter, Torrance's discussion furnishes us with a theological apparatus through which to bring to light the value of liturgical spatio-temporality.
113 Torrance, *Space, Time and Incarnation*, 24, 55.
114 Joris Geldhof makes a similar claim based on the doctrine of incarnation: 'The Son of God took on human flesh at a determinate point of time and located in a certain cultural environment. Likewise liturgy takes on the shape of rituals, prayers, gatherings, and customs of people, always situated in a concrete time and space. Put differently, incarnation is at the root of the sacramental and ecclesial reality of the Body of Christ.' Joris Geldhof, "Penetration–Permeation–Fermentation: Ponderings on the Being of Liturgy and Its Memorial Modes," *Studia Liturgica* 50, no. 1 (2020): 10.
115 Building on the doctrine of incarnation, Ratzinger extensively discusses the liturgy's relation to space and time; Ratzinger, *The Spirit of the Liturgy*, 53–111.
116 Katherine G. Schmidt, *Virtual Communion: Theology of the Internet and the Catholic Sacramental Imagination* (Lanham: Lexington Books, 2020).
117 A typical statement made by Schmidt is as follows: 'The sacraments are a symbolic system of actions, words, and objects that embody both presence and absence. Their corporality bespeaks the grace of God in Christ, who takes flesh to redeem us. And their corporality bespeaks God's absence from us by pointing us back to the mediation that marks our humanity and our distance from God this side of the eschaton. It is precisely here that the sacraments reveal themselves as virtual: they are suspended in the productive and uncomfortable space between presence and absence.' Ibid., 139.
118 https://www.vrchurch.org
119 Indeed, there are controversies around the idea of digital church; see, for example, Alexander Chow and Jonas Kurlberg, "Two or Three Gathered Online: Asian and European Responses to COVID-19 and the Digital Church," *Studies in World Christianity* 26, no. 3 (2020): 298–318; Johnson, "Online Communion, Christian Community, and Receptive Ecumenism: A Holy Week Ethnography during COVID-19," 188–210.
120 Harris Wiseman, "Meaning and Embodiment in Ritual Practice," *Zygon: Journal of Religion and Science* 57, no. 3 (2022): 785.

4 Shall We Blame or Commend AI?

Artificial Moral Agency and AI's Moral Status

Coupled with the doctrine of the *imago Dei*, as demonstrated in Chapter 1, the archetype–ectype thinking indicates that religion takes precedence over morality in spelling out the meaning of human nature. Having examined the idea of artificial intelligence (AI) as the ectype of humanity in relation to AI-powered liturgical enactment, I will probe in this chapter how the notion of AI as the *imago hominis* can advance current studies on AI ethics.

The subject matter of this chapter is 'artificial moral agent' (AMA). The idea of AMA is inextricably tangled with the attempt made to describe computational artefacts and smart machines as moral agents. Given the wide application of AI to smart machines and digital devices nowadays, I will use 'computational artefact,' 'smart machine,' and 'AI' interchangeably in this chapter. The main research question to be addressed is: In what sense, if any, does AI or an AI robot qualify as a moral agent? The questions about AI's moral agency are principally ignited by some moral formulations used widely in AI literature. The concept of agent has occupied a central place in the twenty-first-century AI research. Stuart Russel observes that modern AI is interdisciplinary and connected to many fields (e.g., economy) through 'the tools of probability theory and utility theory.'[1] He continues: 'The central concept of modern AI is the intelligent agent—something that perceives and acts. The agent is a process occurring over time, in the sense that a stream of perceptual inputs is converted into a stream of actions.'[2] Modern AI seems to endow itself with agenthood or make the notion of AMA acceptable by virtue of the practical benefits brought forth by the extensive applications of AI to human lives.

With this being said, the notion of AMA cannot be taken for granted. The terms 'moral agency' and 'moral agent' have a weighty tradition and a ubiquity in usage such that they have become clichés and are understood in different ways. For this reason and for the purpose of this chapter, I will go into details of 'moral agent' and 'moral agency,' respectively in relation to human beings and AI. It will be demonstrated that the archetype–ectype thinking, which carries a strong ontological implication, will serve as the conceptual apparatus through which to flesh out both AI's moral status and the inextricable relationship between human moral agents (HMAs) and AMAs. Based

DOI: 10.4324/9781003356738-5

on the idea of AI as the ectype of humanity, I will argue that AI is embedded with partial moral agency in comparison with human moral agency. This partial moral agency is the extension of human moral agency and explains how AI is applied to human lives while maintaining both the moral significance of human psychosomatic unity and the HMA's moral responsibility. From this vantage point, it can be argued that AI as the *imago hominis* participates in the ethical performance made by the *imago Dei* towards others. From this it follows that the HMA should be willing to accept responsibility for AI's impacts and actions.

In the subsequent text, I will first explore various approaches to defining AMA, with particular attention to Luciano Floridi and Jeff Sanders's endorsement of AMA through exploration of the connection made by them between ethics and the Method of Abstraction (MoA), a mathematical method taken from computer science. Following this, I will examine three critical appraisals of Floridi and Sanders's methodology. Next, the idea of AI as the ectype of humanity will be employed to account for AI's moral agency as partial. In this connection, I will flesh out the ectypal sense of partial AMA through the lens of the moral importance of both the human body and the mind. Finally, I will demonstrate how the idea of partial artificial moral agency as ectypal contributes to recent discussion over responsible AI, engaging theologically with the Edinburgh Declaration on Responsibility for Responsible AI.

4.1 Approaches to Artificial Moral Agent

4.1.1 Moral Agent: Definitions

It is hard to unanimously define moral agency or moral agent. Some philosophers and ethicists strictly confine moral agency to human beings. A prominent representative of these thinkers is Immanuel Kant. In his position, moral agents are rational beings who are moral legislators and autonomously act in accordance with moral laws.[3] In distinction to these thinkers, others construe moral agency and moral agent in a broader sense. For example, Gary Watson elucidates that an agent refers to 'a creature that makes things happen. A creature is an agent insofar as its movements are attributable to it and not just to some subpart of it.'[4] With this in mind, he focuses attention on the concept of HMA, arguing that a moral agent is 'an individual who can, to a significant extent, act effectively and competently in moral matters.'[5] Yet notwithstanding this, Watson's elaboration on moral agency stretches the notion of moral agent to include nonhuman creatures.

> The status of moral agency is bi-directional. Autonomy certifies what an individual with this status can expect (indeed demand) from others (the individual's rights and entitlements); in the other direction, accountability concerns what can be expected (or demanded) of moral agents by others (the individual's duties and responsibilities).[6]

Accordingly, autonomous and accountable creatures qualify as moral agents, though their autonomy and accountability vary in degree.

A similar and more general account of moral agent comes from *The Routledge Encyclopedia of Philosophy*. Therein, Vinit Haksar aims to set forth a generic idea of moral agent:

> Moral agents are those agents expected to meet the demands of morality. Not all agents are moral agents. Young children and animals, being capable of performing actions, may be agents in the way that stones, plants and cars are not. But though they are agents they are not automatically considered moral agents. For a moral agent must also be capable of conforming to at least some of the demands of morality.[7]

This passage points us to the three features that need to be mulled over before defining 'moral agency' and 'moral agent.' First, the expectation of the moral agent shows that the one who expects knows that the agent has the potential to meet moral demands. In this sense, the moral status of an agent is, to a certain extent, presupposed. Second, the agent's moral agency is determined by moral demands, which means that the agent is morally responsible to meet these demands. Although moral demands vary in different communities and across time, it holds true that the moral demands *per se* make agents responsible for their own actions. The presupposed moral statuses of agents are tied up with their responsibility to act in light of moral demands. Third, moral agency reflects a capability to act in a moral manner in order to meet these demands. Moral agency does not mean that the agent is capable of acting in accordance with all moral demands. Agents would fail to live up to some moral demands, and this is their in*capability* to act morally. In this sense, a judgement on moral agenthood is made in accordance with the actualisation of the capacity to meet moral demands.

Leaving aside the jargons such as autonomy and accountability, Haksar's generic definitions of 'moral agent' and 'moral agency' can be stretched extensively not only to speak of humans and animals but also to characterise smart machines and computational artefacts, AI *par excellence*. Whatever moral demands are, computational artefacts are expected to meet these demands, and many believe that AI are capable of acting morally.

The idea that computational artefacts qualify as agents is not novel. More than two decades ago, Ian Kerr suggested that computational artefacts and systems can be considered agents in a legal sense in electronic commerce. Despite that computational artefacts are not legal persons, they can be considered intermediaries rather than mere instruments so that the law of intermediaries enacted for electronic commerce can be expanded to include both human and intelligent intermediaries.[8] In similar manner, a recent report titled 'Privacy, Agency and Trust in Human-AI Ecosystems', published by the Alan Turing Institute, puts spotlight on the attribution of agenthood to AI. While AI-powered systems and devices constrain human agency in one way or another (e.g., AI surveillance), James Wright and his colleagues observe, 'AI systems are themselves increasingly being given, in the minds of some, a form of distributed agency—an ability to act with some degree of perceived autonomy. This is clear in the term "AI agent", used to refer to an AI application that acts on a person's behalf in some way.'[9] The wide applications of AI seem to justify its agenthood.

That said, the idea of AMA is not unanimously approved. Having identified computational artefacts as agents in the legal sense, Kerr maintains that these electronic agents are not moral agents.[10] Christian Herzog raises a moderate critique of AMA. He contends that the alleged AMA is as yet premature. For this reason, the application of AMA to human life 'will potentially negatively feed back on the human social system and on human moral thought itself and its value,' such as diminishing human moral deliberation while making decisions.[11] Indeed, Herzog is concerned about the real-life use of AMAs, but this practical concern also implies that AMA is still a premature idea.

In contradistinction to the disapproval of AMA, some AI philosophers and ethicists take pain to demonstrate how AI qualifies as moral agent. Despite their slightly different understandings of 'moral agency' and 'moral agent', these scholars alike are motivated to delineate AI's agenthood by referencing or comparing it with human agenthood.

4.1.2 The Conceptual Variants of Artificial Moral Agent

The question of AMA has sparked wide-ranging debates among scholars, and some conceptual variants of AMA are formulated to illuminate AI's moral status. In present literature on AI ethics, James Moor's work presents us with a convincing taxonomy of AMA theories.

Moore classifies AMA theories into four categories: ethical impact agents, implicit ethical agents, explicit ethical agents, and full ethical agents.[12] The theory of AMA as ethical impact agents means AI's operation can bring forth ethical consequences and exert a moral bearing on human life. The AI-powered tone detector falls into this category. Such tone detectors can help us to rephrase texts so as to produce a morally sound piece of work. As the implicit ethical agent, AI can promote ethical behaviour and deal with ethical concerns through programming. The studies on ethical guidelines for algorithm design showcase efforts in creating implicit ethical agents. Explicit ethical agents refer to AI-powered artefacts that can autonomously identify ethical information and work out resolutions to the application of ethical principles across various circumstances. AI as a full ethical agent operates in the same way as explicit ethical agents but, importantly, possesses human-level metaphysical features such as consciousness, intention, and free will.

The AMAs in the sense of the first two categories have already been created. Despite his openness to the possibility of building those metaphysical features into smart machines in the future, Moor suggests that the idea of explicit ethical agents can be considered 'the paradigm target example.'[13] His taxonomy helpfully portrays the landscape of current studies on AMA, and his appeal for explicit ethical agents reflects ethical concerns arising in many works about AI ethics. However, Moor is oblivious to the fact that studies on explicit ethical agents are often unwittingly caught up in the issues surrounding full ethical agents—that is, whether or not AI has human metaphysical features. Inquiry into explicit ethical agents generally ends up in exploration of full ethical agents. For example, while examining a robot as a moral agent, John Sullins argues:

> [T]he machine's programmers are not to be seen as the only locus of moral agency in robots. This leaves the robot itself as a possible location for a certain amount of moral agency. Because moral agency is found in a web of relations, other agents such as the programmers, builders, and marketers of the machines, other robotic and software agents, and the users of these machines all form a community of interaction.[14]

Following this, Sullins discusses the three criteria for creating full moral agents: smart machines are fully moral agents when they perform human-level duties, are autonomous and intentional, and fully understand their responsibilities while performing their duties.[15] Hence, he argues elsewhere that the possibility of AMA forces us to broaden the conventional notion of moral agent to include smart machines.[16] Sullins's study proves that the questions about full moral agents are inextricably linked to a thorough investigation of explicit moral agents.

Luciano Floridi and Jeff Sanders's work shows a typical exploration of AMA as full moral agent. Their argumentation begins with an attempt made to redefine the term 'moral agent.' The traditional account of moral agent is, according to them, anthropocentric and individualistic such that ethical discourse often has nothing to do with nonhuman, corporate entities. They instead suggest that artificial artefacts can qualify as moral agents by stretching the meaning of moral agent to include all 'entities that can in principle qualify as sources of moral action.'[17] Hence, moral agents can be both natural and artificial, and there is also natural and artificial evil because the two sorts of moral agents alike bring forth immoral effects.[18]

Floridi and Sanders develop the idea of AMA with recourse to the MoA, which comes from the field of computer science. The MoA adopts the method that 'discrete mathematics is used to specify and analyse the behaviour of information systems.' It seeks to construct a model in which the variables align with observables in reality. A variable, which is common to scientific modelling, refers to a symbol that points to 'an unknown or changeable referent.' When this symbol 'hold[s] only a declared kind of data', it is named typed variable. An observable is composed of a typed variable and the connotation the latter carries of features of the system that it depicts.[19]

Under the auspices of the MoA, one can build and formalise the model of a system, developed at levels of abstraction. Each level of abstraction 'is a finite but non-empty set of observables, which are expected to be the building blocks in a theory characterised by their very choice.'[20] Floridi and Sanders argue that the level of abstraction at which we discuss moral agents largely relies on the conviction that human beings are moral agents. However, this level of abstraction is lower and includes too many details about HMAs. In order to steer clear of the anthropocentrically defined meaning of moral agent, they claim that a higher level of abstraction must be adopted so that fewer details about moral agents need to be considered.

Floridi and Sanders argue that the level of abstraction at which the AMA can be conceived needs to be upgraded by considering another three criteria: interactivity (interaction with environments), autonomy (capability to change state independently), and adaptability (learning to operate in a new way).[21] Machine learning (ML) is cited in support. ML can interact with its environment, is autonomous and non-deterministic, and can learn to change its model of operation to adapt to new circumstances.[22] In light of this upgraded, higher level of abstraction, a moral agent can be defined as an agent that 'is capable of morally qualifiable action' which 'can cause moral good or evil.'[23] In this manner, Floridi and Sanders articulate a notion of full moral agent.

4.1.3 *The Critical Appraisals of Artificial Moral Agents*

By identifying computational artefacts as moral agents, Floridi and Sanders rightly recognise the moral importance of these artefacts for human life. By the same token, AI has a tremendous bearing on humans and changes the way of human life. Yet their methodology of recasting the concept of moral agent invites criticism.

First, Floridi and Sanders's methodology presupposes the computerisation of morality. Taking the MoA from computer science, they implicitly equate the essence of morality with information processing. They believe that moral observables can unequivocally and forthrightly reveal moral nature, *and* that moral models can be built every bit as similarly as scientific models. This is reminiscent of the computational model of consciousness (see Chapter 2, subsection 2.1.2). Both the computational model and the MoA are utilised to deploy digital technology to address issues beyond the sphere of technology. However, the essential difference between morality and science falls through the cracks. Morality is complex, and moral rules that underlie moral observables may vary across time. Hence, computerising or modelling morality at the levels of abstraction is nothing other than simplifying moral life.[24]

Second, Floridi and Sanders's redefinition of moral agent is a non-starter since they make one particular level of abstraction dominant among others and blur the boundaries between levels of abstraction by adopting univocal senses of the criteria for formalising models. As noted earlier, they draw on

ML and highlight its interactivity, autonomy, and adaptability as the three criteria for justifying the idea of AMA. Yet, they are oblivious to the distinction between the meanings of these criteria when understanding them in terms of AMA and HMA, respectively. At one level of abstraction, ML is interactive, autonomous, and adaptive. Be that as it may, behind the scenes of these criteria is the process of design. AI programmers, designers, stakeholders, and end users involve in the creation of AI. (This corollary coincides with my exploration of ML in relation to liturgical enactment in Chapter 3, subsection 3.3.2.2.) That is to say, different judgements about these criteria arise because the criteria can be formed from the respective perspectives of AI designers and users. As Frances Grodzinsky and his colleagues note, despite that computational artefacts could be viewed as moral agents at the user's level of abstraction, they are never deemed to have moral agency at the designer's level of abstraction, and so designers always take on moral responsibility.[25] (We will have a closer look at moral responsibility and AMA in the Section 4.3.)

Floridi and Sanders's computerisation of morality under the auspices of the MoA is not novel. Their methodology is similar to Blay Whitby's computing morality. Whitby suggests that AI can replace humans as moral decision-makers under certain circumstances by virtue of its computing and autonomous system.[26] However, other scholars level criticism against such computerisation. Allen Newell argued back in 1980 during his presidential address to the American Association for Artificial Intelligence that computer systems only provide a description of reality at a certain level, which means that this description is not working at all levels. In his position, 'computer system levels are a reflection of the nature of the physical world. They are not just a point of view that exists solely in the eye of the beholder. This reality comes from computer system levels being genuine specializations, rather than being just abstractions that can be applied uniformly.'[27]

Some counterarguments have recently been raised against Floridi and Sanders's methodology, taking issue with the homogenisation of AMAs and HMAs. Here I will concentrate on three counterarguments: (1) Deborah Johnson and her colleagues' distinction between intendings and intentionality of moral behaviour, along with their analysis of the social nature of moral behaviour; (2) Robert Sparrow's differentiation between scientific and moral dilemmas; and (3) Steve Torrance's emphasis on the organic underpinning of morality.

4.1.3.1 *Intentionality Versus Intendings*

Deborah Johnson submits that the MoA fails to capture the full sense of moral agency. All technologies, including computation systems, 'are produced, distributed, and used by people engaged in social practices and meaningful pursuits', but certain levels of abstraction have nothing to do with human social practices.[28] By way of illustration, Johnson and Keith Miller compare the human person's and the machine's actions of opening the door. Both a human person and an electronic locking system with an electric eye can open the door for someone who is carrying a large and heavy package. These two actions of opening the door are different by their nature: 'The function performed is equivalent, but the underlying processes (voluntary, autonomous act versus mechanical operations) are significantly different.'[29] The human person's action *for* other human persons is embedded with social meaning, but the electronic locking system's operation *per se* is not social, though the mechanical operation can be viewed as the extension of human social practices. We will return to this point later while elaborating on ectypal artificial moral agency. Theologically speaking, this extension of human social practices is indispensable for the *imago hominis*.

Johnson and Miller stress that '[a]bstractions are the work of humans and the abstractions themselves do not exist separately from humans.' It is specious to characterise computational artefacts as having moral agency by virtue of one particular level of abstraction.[30] Computational artefacts cannot operate completely independently of humans and should not be considered full moral agents. To flesh out the difference between the AMA and the HMA, Johnson distinguishes between intentionality and intendings to act: 'While computer systems do not have intendings to act, they do have intentionality.'[31] She continues:

> Computer systems and other artifacts have intentionality, the intentionality put into them by the intentional acts of their designers. The intentionality of artifacts is related to their functionality. Computer systems (like other artifacts) are poised to behave in certain ways in response to input. ... The output (the resulting behavior) is a function of how the system has been designed and the input I gave it. The system designer designed the system to receive input of a certain kind and transform that input into output of a particular kind, though the programmer did not have to specify every particular output for every possible input.[32]

The intentionality of artefacts rests with the human designer's intentionality, which is actualised through intendings to act. At the same time, this inbuilt intentionality of computational artefacts means that they can, to a certain degree, operate independently of humans, which goes some way to giving one the impression that computational artefacts are full moral agents.

Johnson's distinction between intendings and intentionality rules out the possibility to attribute the HMA's psychic features (e.g., consciousness) to AMAs. Computational artefacts do not have intendings to act precisely because intendings to act arise from freedom. 'The intending to act is the locus of freedom; it explains how two agents with the same desires and beliefs may behave differently.'[33] Computational artefacts are designed and produced in a standardised way and are expected to operate in a specific way. As such, computational artefacts can never be full moral agents since they cannot be completely detached from the human designer's intendings to act.

With that said, Johnson categorically refuses to deny the moral status of computational artefacts and holds the view that computational artefacts belong to the human moral world.

> Computer systems (and other artifacts) can be part of the moral agency of humans insofar as they provide efficacy to human moral agents and insofar as they can be the result of human moral agency. In this sense, computer systems can be *moral entities but not alone moral agents*. The intentionality and efficacy of computer systems make many human actions possible and make others easier and therefore more likely to be performed.[34]

Johnson elucidates this viewpoint with her colleagues in two specific respects. First, that computational artefacts are part of human moral agency means that they should always be 'conceptually tethered to human agents' in such a sense that it is humans who create, design, and use computational artefacts.[35] Second, computational artefacts as moral entities have surrogate agency. Surrogate agents are employed to perform tasks on behalf, and in the interests, of users. Human surrogate agents and computational surrogate agents differ in that the former 'have a first-person perspective independent of their surrogacy role', whereas computational artefacts 'do not have interests, properly speaking, nor do they have a self or a sense of self.'[36] Hence, as surrogate moral agents, computational artefacts only pursue the interests of their human users. As a result, humans should constantly take on moral responsibility for the consequences of their artificial surrogate agents.

4.1.3.2 The Scientific Versus the Moral

Robert Sparrow raises a second argument against Floridi and Sanders's viewpoint of AMA. Unlike Johnson and her colleagues who directly targets the MoA, Sparrow draws the nature of ethics to the discussion table regarding AI's moral agency.[37] He argues that all endorsement of the possibility of moral smart machines is rooted in an inadequate understanding of the nature of ethics.

Sparrow turns our attention to the personal nature of morality: 'ethical dilemmas attach to agents in such a way that they are essentially dilemmas for particular people.'[38] Computational artefacts designed and produced in a standardised way are unable to deal with all specific, personal moral dilemmas. For this reason, Robert Sparrow and Linda Sparrow maintain that it is unethical to place robots as the substitution for humans in aged care.[39] The personal nature of ethics cannot be reproduced across circumstances for all people.

Having brought to light the personal nature of ethics, Sparrow explores the impossibility of AMA through a methodological lens, that is, the radical difference between the methods used to address scientific and moral dilemmas.

> Scientific questions are objective in the familiar sense that the true value of scientific claims does not depend on who is making them. This means that such questions are fundamentally impersonal.... The phenomenology—and also the logic—of ethical questions suggests that they are also objective: a person facing an ethical dilemma cannot make their ultimate decision correct simply by approving of it. However, in contrast with scientific dilemmas, ethical decisions are tied to particular people—they are decisions for them in a non-contingent sense. That is to say, ethical dilemmas are fundamentally personal.[40]

Sparrow makes it clear that while both science and ethics share objectivity, moral dilemmas and decisions are primarily characterised by subjectivity, which means that morality is, by its nature, personal. Following this line of thought, it can be argued that variables for scientific modelling are unable to connote the human agent's handling of moral issues according to moral values held by individual human persons. Jobst Landgrebe and Barry Smith rightly note, 'Values are not variables that can be put into a mathematical model in order to be multiplied by weight parameters and summed over.'[41] In short, morally relevant features cannot be thoroughly programmed into these systems through scientific modelling.

Sparrow's distinction between the moral and the scientific is echoed in other studies. John-Stewart Gordon argues that the idea of AMA falls into an ethical pitfall—that is, without paying due attention to pluralism in ethics.[42] While there is no single ethical theory that can be used to address moral issues universally, AMAs are expected to deal with moral dilemmas across cultural settings in accordance with certain moral norms built in with *fixed* algorithms. As a result, the particularities of each human person's approach to handling moral issues are discounted.

This scientific-moral distinction, along with the personal nature of ethics, entails two implications.

> First ... it means that we cannot escape responsibility for making ethical decisions by acting on the advice of others. Second, the character, and the life history, of the individual facing the dilemma may enter into our account of their reasoning about the dilemma and thus, to a certain extent, into our account of the nature of the dilemma.[43]

Since AMAs are incapable of replacing humans, it is always HMAs who make decisions and perform moral behaviours. All the more important is that the HMA's personal moral particularities—which cannot be computerised through the MoA—should always be reckoned with while applying AMAs to human lives.

4.1.3.3 The Organic Underpinning of Morality

While developing his counterargument against the idea of AMA, Steve Torrance places emphasis on the inextricable relationship between humans *qua* biological organisms and humans *qua* moral agents. His point of departure is the basic principle that the mind plays a significant role in human morality. As early as 1986, Torrance wrote that 'the notion of mind is central to our ethical thinking.'[44] From this it follows that the possibility of AMA is contingent upon the reproduction of the human mind in AI.

Torrance's attention to the interconnection between the mind and morality is not confined to human mental activities. Rather, starting from the mind, his investigation of AMA concentrates on humans as biological, psychosomatic organisms. The organic underpinning of human morality consists of five aspects.[45] First, biological organisms and machines radically differ from each other. Second, due to this organism-machine dichotomy, machines can never be intrinsically moral. Third, human biological conditions serve as the *organic* foundation for the emergence of moral thinking, feeling, and behaviour. Fourth, sentient or phenomenally conscious agents can either have moral concerns or make moral appraisals. Fifth, genuine sentience and consciousness are unique to biological organisms.

Torrance's elaboration on the organic underpinning of human morality is in line with what I have discussed in Chapter 2 about the close link between human biological conditions and consciousness. (We will return to detail this point in the following section while developing an ectypal account of AMA.) Here Torrance's organic view of morality foregrounds a major reason for the impossibility of the computerisation of ethics. As I have reiterated, humans as psychosomatic beings cannot be emulated in silicon-based AI. This point is clearly present in the below block quote from Torrance's work, showing the concatenation of human biological bodies, minds, and morality.

> Morality is fundamentally a domain of creatures that have an internally organized existence rather than an externally organized one—that is, creatures that exist not merely as artefacts whose components have been assembled together by external designers, but which exist in a stronger, more autonomous sense. On this view the moral universe can contain only, or primarily, *self-organizing, self-maintaining entities, that have an inherent, active telos to survive and thrive*—entities whose sole current instances are natural biological organisms. Moreover, consciousness is an emergent property of the physiological structures of these naturally occurring self-maintaining organisms. So, in this view, both moral normativity—the expectations and requirements that we take to be central to ethics—and *the facts of lived, sentient experience, are seen as integrally bound up with the forms of self-organization and self-maintenance, which are essential features of biological organisms.*[46]

It is worth clarifying that Torrance's purpose is not to biologise moral agency, nor does he attempt to render moral agency *solely* dependent upon human biological particularities.[47] Moreover, he does not seek to rupture the moral connection between HMAs and AMAs due to AI's lack of biological features. Rather, Torrance's main purpose is to make the point that the biological features of the HMA *qua* organism entail *unique* implications for moral agency, which is not susceptible to the design of digital technologies. Contrary to Floridi and Sanders's position, the 'self' in self-organising and self-maintaining cannot be utterly computerised at any level of abstraction. There are no algorithms or mathematical models that can reproduce organic systems for moral agency.

Torrance's argumentation from the mind to human biological features is resonant with the holistic reading of the *imago Dei*—that is, the human being as a psychosomatically united whole. This point of view underlies the conceptual connection between AI as the ectypal of humanity and AMA. I will go into comprehensive details of this point in the next section. Furthermore, Torrance's view of the organic underpinning of morality reminds us of the distinction between emulation and simulation, as elucidated in Chapter 1 subsection 1.4: 'Simulation … attempts to mimic; emulation attempts to reproduce.'[48] Floridi and Sanders's deployment of the MoA into the computerisation of morality is, *de facto*, to simulate HMAs while creating AMAs; no emulation takes place during the creation of AMAs due to the impossibility of the mathematisation of human organic features through algorithms.

4.1.4 *Summary*

I have thus far examined the conceptual variants of AMA, with a focus on Floridi and Sanders's MoA and computerisation of ethics. It has been demonstrated that mathematical models and algorithms cannot reproduce full moral agenthood in AI. Deborah Johnson, Robert Sparrow, and Steve Torrance together showcase that the HMA as a psychosomatic organism performs moral agency in a unique way compared to the AMA.

It is misleading to suggest, however, that AI is a value-free instrument and has nothing to do with human morality. Although Floridi and Sanders's methodology is cast into doubt, their intention to demonstrate AI's moral status is commendable. As Mark Coeckelbergh rightly notes, 'To the extent that we become dependent on AI, AI is more than a tool we use; it becomes part of how we are, and how we are at risk, in the world.'[49] Indeed, AI is also part of human moral life. But in what sense? In the next section, I will leverage the idea of AI as the ectype of humanity to account for the meaning of AMA.

4.2 An Ectypal Account of Partial Moral Agency

4.2.1 *Partial Moral Agency as Ectypal*

Deborah Johnson, Robert Sparrow, and Steve Torrance take pains to hedge morality against computerisation insofar as morality is more complex than constructing mathematical models. Furthermore, both their positions and

Floridi and Sanders's argument caution us against the view that computational artefacts are morally neutral. AI's design and applications together prove that AI-powered devices are laden with moral values, which are built in AI systems through algorithms, data collection (as discussed in Chapter 3), users's utilitarian demands, and so forth. It suffices to say that AMAs and HMAs are inextricably associated.

However, two questions remain to be addressed. (1) In what metaphysical sense shall we understand the tether or connection between computational artefacts and humans? This question concerns the metaphysical foundation for drawing the distinction between AMAs and HMAs. At the same time, the first question drives us to explore how the *technologically built* connection between AMAs and HMAs resonates with a metaphysical account. (2) Is it possible to understand the connection between AMAs and HMAs in a non-utilitarian sense? Indeed, humans are morally intertwined with computational artefacts while designing and using artefacts for utilitarian purposes, that is, to borrow Johnson's language, human intendings to act through artefacts. In this light, utilitarian purposes often serve as the *only* hermeneutical tool for the interpretation of the connection between HMAs and AMAs, resulting in the eclipse of the ontological connection between what the HMA is and what the AMA is.

Mahi Hardalupas puts forward the idea of partial moral agency, which keeps the close ties between HMAs and AMAs while differentiating these two kinds of moral agenthood. She suggests that there are four conditions for judging full moral agenthood: (1) action evaluated by moral rules; (2) acting according to moral rules; (3) possibility to follow different rules; (4) moral motivators, which mean either believing an action as moral or the rules to follow as moral.[50] Smart machines and computational artefacts are presently partial moral agents because they can only fulfil certain aspects of these four conditions, especially the first three. It is unclear whether Hardalupas's four conditions would eventually result in a rule-based morality, a variant of computerisation of morality. Neither does Hardalupas unfold the four conditions or discuss whether humans can create smart machines that are able to fulfil all of these conditions in the future. In line with Torrance's emphasis on the role of the mind in morality, recent studies on AI ethics foreground the impossibility of creating AI that is capable for moral motivation.[51] According to the findings of the previous chapters, artificial consciousness can never become a human-level one, and so the AMA will remain to be partial moral agent.

With this in mind, the idea of partial moral agency is a preferred conceptual apparatus through which to shed light on the fact that computational artefacts are part of human moral agency. It would be helpful to compare 'partial moral agency' with Johnson's 'surrogate agency' in this context. By claiming that computational artefacts as surrogate moral agents only pursue the interests of their human users, Johnson places emphasis more on the separation than on the connection between AMAs and HMAs. By contrast, 'partial moral agency' intensifies that AMAs differ radically from HMAs and yet are part of human moral agency. Furthermore, compared with 'surrogate agency,' the notion of partial moral agency always implies and presupposes full moral agents, namely, HMAs. As such, the HMA should be willing to accept moral responsibility for day-to-day AI applications. I proceed to expand on the concept of partial moral agency from the perspective of the theological ontology of archetype–ectype.

We need to recall the key components of the idea of AI as the ectype of humanity, as discussed in Chapter 1, in order to flesh out the theological meaning of partial moral agency. The archetype–ectype thinking is teamed up with the doctrine of the *imago Dei*, which is in turn concomitant with the doctrine of creation, particularly with the idea of *creatio ex nihilo*. As such, the thinking of archetype–ectype carries both ontological and ethical implications. As the ectype of God, the human being is not on a par with God; rather, there is an ontological chasm between the divine archetype and the human ectype, which connotes the essential differences between the creative activities of God and humans—that is, God creates out of nothing, but humans create out of something. The qualitative distinction between divine creation and human creation turns out that there is an essential difference between humans as the consequence of God's creation and computational artefacts as the consequence of human creation. Furthermore, as God's ectype, human beings should continue to become moral in order that they can correspond in all parts to God by displaying God's attributes. In this vein, the ontological chasm between the archetype and the ectype is intertwined with their moral connection and resemblance. Likewise, human action, including human creation in an ectypal sense, bears moral implications. Hence, the HMA as God's creature and the AMA as artefact differ from yet connect closely with each other by their natures.

This recollection of the idea of AI as the ectype of humanity showcases that both the idea of AMA and the archetype–ectype thinking emphasise the intrinsic link between what the human being is and what the human being acts. They differ in that the archetype–ectype thinking theologically grounds this intrinsic link in God and his creative work, whereas the idea of AMA portrays this intrinsic link from the perspectives of technology philosophy and ethics. All the more salient is that AI as the ectype of humanity sits well with the notion of partial moral agency. *A bold claim can be made that the theology of archetype–ectype can lay an ontological and moral foundation on which partial moral agency can be conceived of and attributed to computational artefacts.*

Taking together the key components of the idea of AI as the ectype of humanity, the theological account of partial moral agency—that is, the notion of ectypal partial moral agency—can be developed according to the following two syllogisms. The first syllogism is related to the difference between divine and human creation and to the distinction between AMAs and HMAs:

1 the major premise: human moral agency is the consequence of God's creation;
2 the minor premise: the moral agency of artefacts is the consequence of human creative work;
3 the conclusion: artificial moral agency differs from human moral agency due to the essential difference between divine and human creative work.

This syllogism takes issue with Floridi and Sanders' computerisation of morality through modelling and the MoA in that the latter methodology is rooted in the conviction that AMAs equate to HMAs, which, in turn, makes human creative work tantamount to God's *creatio ex nihilo*. Hence, the theological idea of AI as the ectype of humanity joins the ranks of Johnson, Sparrow, and Torrance in disapproval of AMA as full moral agent.

The second syllogism is derived from human creative work and its moral significance in relation to God's creative work:

1 the major premise: The human being is the ectype of God and thus imitates God's creation;
2 the minor premise: God's creation is coupled with the mediation of morality to the human being as his ectype;
3 the conclusion: The human creation of computational artefacts is concomitant with the mediation of morality.

This syllogism tallies with Johnson's argument that AMAs should conceptually be tethered to HMAs. Yet, it is beyond doubt that this syllogism expands and enriches the meaning of 'tethered'. The mediation of morality in the human creation of computational artefacts conveys a more dynamic rather than mechanic connection between AMAs and HMAs. At the same time, this syllogism is not content with the claims that the AMA is a surrogate agent (Johnson) or that humans perform moral actions through AI (Sparrow and Torrance). Rather, the AMA mediates human morality.

In light of these two syllogisms, we can unpack the idea of partial artificial moral agency in three aspects. First, partial artificial moral agency is predicated upon the fact that the computational artefact is the ectype of humanity. The meaning of ectype epitomises how computational artefacts take shape in the human mind. As noted at the beginning of this study, human beings, AI researchers in particular, will explicitly or implicitly serve as models while designing AI. Human designers mediate their moral values into AMAs while creating them. Understanding AMAs in the ectypal sense indicates that computational artefacts are not merely part of human moral agency but also the extensions of human morality. It is in the sense of extension that human–machine relationships can be properly understood.

For example, the desire for artificial companions, at bottom, exhibits the lack of human companions. (We will probe this point in Chapter 5). Seen from this perspective, the theological account of the extension of human moral agency in AMAs demonstrates that AMAs belong into the moral sphere of HMAs, albeit the ontological chasm between them.

Second, artificial moral agency is partial because it is ectypal, limited, and consequently only related to particular moral issues. A clarification needs to be made here. Conjoining the ectypal and the partial (limited) does not imply that humans as the ectype of God possess only partial rather than full moral agency. Theologically speaking, the idea of full artificial moral agency implies that human creation is on a par with God's creation out of nothing, in which human full moral agency originates. Artificial moral agency as partial shows that humans cannot *fully* mediate their moral agency to computational artefacts in their creative work in the same way as God did with HMAs in light of archetype–ectype thinking. Partial AMA reveals the limitations of human creative work. As will be unfolded in the subsections 4.2.2 and 4.2.3, these limitations range over both biological and mental factors.

Third, the concept of partial artificial moral agency as ectypal implies that it is always the HMA who is held responsible for the ethical decisions and behaviour performed by AMAs on the ground of the ontological connection between the archetype and the ectype. The HMA cannot escape moral responsibility after deploying computational artefacts to make ethical decisions. It is always the HMA who designs and uses computational artefacts to address ethical dilemmas. We will probe the concept of responsible AI through the angle of ectypal partial artificial moral agency in Section 4.3.

This triple moral implication, derived from the above two syllogisms, shows that the principle that the AMA has ectypal partial moral agency opens up a theological way to deal with questions related to AI's moral status. Now I move on to spell out the theological meaning of 'ectypal' in 'ectypal partial moral agency' in view of the HMA as the psychosomatic being.

4.2.2 *Ectypal Partial Moral Agency: Morality and Human Bodies*

Theologically speaking, the meaning of ectypal partial moral agency is associated with the HMA's biological body as the consequence of God's creation. Christianity binds morality and human bodies together from its inception. Paul the apostle asked Christians in Rome to present their bodies 'to God as instruments of righteousness' (Rom. 6: 12-13; NRSV). Augustine maintained that the mortality of human bodies is tangled up with original sin, which resulted in the concupiscence of the flesh and voluntary immoral action.[52] As David Hunter notes, 'the body is [for Augustine] the source of the evils experienced by the soul.'[53] Contemporary Christian ethicists also pay great attention to human biological features.

Jürgen Moltmann contests that inasmuch as the concept of life must take in all relevant biological findings, human moral life is tied up with their biological bodies.[54] Victor Austin suggests that human bodies are required for the fulfilment of the mutual ethical relationship between humans, i.e., friendship, which is ultimately grounded in human friendship with God.[55] A clarification needs to be made here: attending to human bodies does not mean that Christian ethics is inclined to biologise morality.[56] Rather, the purpose of Christian ethics is to emphasise the indelible role of human bodies in morality.

The moral importance of the human body demonstrated in Christian ethics shows that the ectypal implication of partial moral agency partly arises out of the close tie between HMAs and their bodies. This theological understanding of ectypal AMA is in agreement with recent studies on AI ethics. For example, Wendell Wallach and Shannon Vallor argue that human embodiment and bodily interaction with the moral world are indispensable for the human cultivation of virtue.[57]

> While many, if not most humans fail to cultivate and fully refine their moral capacities, our species has evolved the potential for individuals to develop advanced moral competencies or virtues through a broad range of parallel but massively integrated systems for moral understanding, perception, emotion, reflection, and imagination. Such systems recruit cognitive, sensorimotor, and affective capacities in the human body… Thus to accomplish [artificial systems] on a circuit board, *without the benefit of the vast sensitivities of the exteriorized animal body*, is no great miracle.[58]

Wallach and Vallor underline the importance of the human body for morality while maintaining the unity of the mind and the body. Theoretically speaking, they are open to the possibility of AI's fully embodied virtue in the future. Yet, the design of such AI systems must start from scratch with the concept of embodied virtue as the point of departure.[59] The HMA's embodied virtue serves as a benchmark for conceptualising the possibility of the AMA being imbued with fully embodied virtue. Coupled with the idea of AI as the ectype of humanity, Wallach and Vallor's openness to the possibility of AI's fully embodied virtue does, in fact, endorse AI's ectypal partial moral agency in light of the moral significance of the human body. At the same time, their study foregrounds a critical limitation of AMA: artificial moral agency is limited and related to particular moral issues due to the AMA's incapability for the cultivation of virtues through human-level embodied bodies.

It is, however, worth noting that the above survey of the body-morality relationship in the history of Christian ethics illuminates that the moral significance of the human body is not limited to the human embodiment. That is to say, even if AI robots have the same embodiment as humans, the AMA is still partial and ectypal insofar as the human body is more than the physical form. Owen Holland, Rob Knight, and Richard Newcombe's observation is on the mark.

> A human body is more than just an external form—its internal structures (bones, joints, muscles, and tendons) seem qualitatively different from those of robots, as do the movements they produce. Our robot, then, should combine a humanoid outer form with human-like internal structures that would have to be controlled in much the same way as the human body is controlled.[60]

In addition to the physical embodiment, the biological features of the human body matter to the distinction between the AMA and the HMA and are of importance for the meaning of *ectypal* partial moral agency. Given the bold claim made by some scholars that artificial neural networks can emulate the human neural network (see Chapter 2, subsection 2.1.2), I proceed to leverage interdisciplinary studies on neuroscience and morality to expand on the meaning of the AMA as ectypal.

Neuroscientific studies have brought to light the close bond between ethics and the neural activities in the brain. An example of such studies is that which demonstrates the ventral and medial aspects of the prefrontal cortex in relation to social emotions (e.g., shame, embarrassment, and envy), which convey moral values.[61] It is, however, worth noting that the studies on the association between neural activities and ethics is not tantamount to the biologisation of morality. Rather, neuroscientific studies explain the biological mechanism within the brain that, together with social and cultural contexts, underpins human moral action. As has laid out in Chapter 2 that artificial neural networks fail to reproduce the networks of neurons within the human brain, it stands to reason that this biological mechanism, along with neural activities thereof, reinforces the distinction between the HMA and the AMA, which aligns with the archetype–ectype thinking.

Patricia Churchland's inquiry into the connection between moral values and neurophysiology as well as neurobiology can add details to the meaning of ectypal artificial moral agency. Churchland maintains that morality is linked with social life and neural mechanisms within the brain.[62] The social nature of morality can be characterised by attachment, which 'refers to the dispositions to extend care to others, to want to be with them, and to be distressed by separation.'[63] All of these dispositions are biologically predicated upon neural circuitry.[64] This is so because social practices (e.g., social cooperation) require learning, which, in turn, brings forth structural changes in the brain during the long course of human evolution.[65] As such, the brain and neural mechanisms serve as a biological foundation for moral progress and the formation of moral values within social groups by addressing social problems. Churchland contends:

> navigation of the social world mostly depends on the same neural mechanisms—motivation and drive, reward and prediction, perception and memory, impulse control and decision-making. These same mechanisms can be used to make physical or social decisions; to build world knowledge or social knowledge, such as who is irascible, or when am I expected to share food or defend the group against intruders or back down in a fight.[66]

Despite that Churchland's intention to naturalise morality and make morality somehow independent of religion will invite much criticism, his endeavour in connecting moral values to neurobiology in light of human social life underwrites the uniqueness of the HMA vis-à-vis the AMA.[67] That is, the body's deep involvement in the formation of moral values over millions of years of evolution cannot be explained away by mathematical models and algorithms created within months or years. At the same time, Churchland's analysis of the association between moral values and neural mechanisms lends scientific support to the Christian ethical account of the moral significance of human biological bodies.

Flowing from the *holistic* reading of the *imago Dei* is the idea of AI as the ectype of humanity, which partly yet significantly accentuates the distinction between AI and humans by virtue of human *biological bodies*. This distinction also puts flesh on the meaning of ectypal AMA. AI's lack of human biological bodies is an ontological component of *ectypal* artificial moral agency.

4.2.3 Ectypal Partial Moral Agency: Morality and the Mind

Placing emphasis on the psychosomatic nature of the human being, the archetype–ectype thinking renders the meaning of *ectypal* artificial moral agency associated with the human mind. In their critical appraisals of Floridi and Sanders's computerisation of ethics, both Deborah Johnson and Steve Torrance draw our attention to how the mind plays a part in human morality. As with them, many AI ethicists and philosophers argue that the human mind flags up the radical difference between the AMA and the HMA.

Richard Spinello's articulation of moral agency refutes Floridi and Sanders's identification of interactivity as an essential criterion of moral agency. Spinello contests that 'mere causal efficacy and interactivity with the environment does not imply freedom and intentionality, which are necessary conditions of moral agency.'[68] It is clear that in Spinello's position freedom and intentionality as the two aspects of the mind are two crucial components of moral agency and distinguish between the AMA and the HMA. For this reason, moral action must be deliberate and purposeful and originate in the human will.[69] Subjectivity and personal experience are, therefore, indispensable for moral action.

> The person knows "I did that," and so this is "my action." The person's awareness that through this choice he is the efficient cause of this action constitutes an "ethical experience" and signals his moral accountability for what has transpired. A computer system cannot have an ethical experience, because it merely manipulates symbols. It does not know the meaning of those symbols and therefore lacks both an awareness and understanding of what it is doing.[70]

He continues:

> Since the action belongs to the self, the self experiences that action as something that it can do but need not do. If a moral agent chooses X, he knows that he could also have chosen Y or simply refused to make any choice. By reflecting on this lived experience, we come to recognize that the human self has this capacity for voluntary action that is directed toward an object and has its beginning in the self, from which it proceeds such that there is no other cause ulterior to the self.[71]

Taken together, the awareness of self, the will, ethical experience, and intention showcase that morality is partly yet significantly engendered through the mind and in consciousness. As such, Spinello maintains that the AMA cannot be a fully human agent.[72]

Spinello's study is in concert with my exploration of qualia and AI in Chapter 2. AI's simulated artificial consciousness falls short of qualia that are the properties of conscious experiences. It is misleading to attribute these mental features to AI so as to justify the full agenthood of AMA. From this vantage point, it can be argued that the ectypal status of AMA is rooted as much in its lack of the moral qualia of consciousness as it is in its lack of human-level biological bodies.

Kenneth Einar Himma presents us with a different approach to AI's moral status, which is conduce to the construal of *ectypal* artificial moral agency. As with Spinello, Himma contests that moral agency is inextricably tied to consciousness. Agency involves with mental states and is consequently intertwined with volitions, beliefs, desires, and intentions.[73] Following this, Himma sets forth four features of moral agency: 'for all X, X is a moral agent if and only if X is (1) an agent having the capacities for (2) making free choices, (3) deliberating about what one ought to do, and (4) understanding and applying moral rules correctly in paradigm cases.'[74] These four features boil down to the assertion that moral agency is bound up with the instantiation of mental states in contexts while performing moral action and addressing moral dilemmas.[75] Himma is open to the possibility of smart machines that are created with these four features in the future.[76]

Yet notwithstanding this, my elaboration on the difference between human and artificial consciousness in Chapter 2 casts the possibility of conscious AMA into doubt. For all that the AMA can operate in a manner of free choice, deliberation, and understanding, it would be naïve to equate these functions with the capacities of the human mind themselves. Thomas Powers and Jean-Gabriel Ganascia's remark on this matter is pertinent: 'since a machine does not determine its own ends or goals of action, but *acts on goals given to it from outside*, invoking will—that is, diving errantly into machine motivations—would seem foolish.'[77] From the AMA's *prima facie* mental functions that are in essence given by the HMA, it follows logically that the ectypal implication of artificial moral agency lies in the AMA's incapacity for the substantiation of all mental states related to moral matters because the purposes of moral actions are personal by their nature. From this it follows that the AMA only performs certain moral actions to address particular moral issues according to the goals established by AI researchers.

Himma's study turns our attention to the nuts and bolts of *ectypal* artificial moral agency. Here I will focus attention on moral decision-making, a volitional action that catches much attention of AI ethicists and philosophers. Theological discussion over moral decision-making can help us discover the sense in which the AMA cannot substantiate moral volitions and fails to substitute the HMA for making moral decisions.

Rowan Williams's detailed treatment of Christian moral decision-making is of help here. As with Spinello's emphasis on the self, he maintains that Christian self-knowledge—that is, 'knowledge of who or what they are'—underlies Christian moral action and decision-making. Following this, he suggests that this self-knowledge is secured 'by meditating on the relations in which [Christians] already stand' and in which each Christian has her own unique place. As such, Christian morality is primarily predicated upon the relations they are involved in rather than upon acquiring more moral knowledge and truth.[78] As will be seen anon, this relational understanding of moral decision-making coincides with the collective implication of the *imago Dei* and the moral significance thereof.

Williams's relational characterisation of Christian morality entails that Christian moral decision-making is tied up with and guided by the relationship with God and communal relationship (i.e., one's relationship with others within faith communities).[79] For this reason, Christian moral decision-making carries a special theological meaning: action following moral decisions is a gift given by God to the community and should thus reflect God's character.[80]

> What will guide me is the need to show in my choices the character of the God who called me and the character of the community I belong to; my God is a God whose concern for all is equal; *my community is one in which all individual actions are measured by how securely they build up a pattern of selfless engagement with the interest of the other* – which in itself…is a manifestation of the completely *costly directedness to the other* that is shown in God's act in Christ.[81]

Owing to her religious belief and communal commitment, a Christian's moral decision-making is concomitant with struggle, reflection, hesitance, obedience within the community, commitments to others, acceptance of one another's moral failures.[82] All of these showcase the role that the human mind plays in moral decision-making.

Williams's analysis of Christian moral decision-making puts flesh on the ectypal meaning of artificial moral agency. For all that he elucidates the meaning of moral decision-making in terms of Christian ethics, the relational nature of decision-making remarkably testifies to the eccentricity of the holistic reading of the *imago Dei*: the meaning of being human is partly predicated upon others (see Chapter 1, Section 1.2). This being so, this relational interpretation of Christian moral decision-making reveals the common features of all moral decision-making. That is, by virtue of AI's pursuit of accuracy, rapidness, efficiency and success in addressing issues, the AMA can make moral decisions in nanoseconds. By contrast, the mental activities concomitant with moral decision-making—such as hesitance, inner struggle, and reflection—cannot be reproduced in the AMA due to AI's simulated rather than emulated consciousness. Furthermore, it is AI researchers who, to borrow Sara Lumbreras's language, 'delegate decisions to' AMAs through algorithms; these decisions can be explained by 'sufficiently detailed instructions' and exhausted by sufficient cases.[83] Seen from this perspective, the idea of *ectypal* artificial moral agency signifies that the AMA is significantly limited in addressing moral issues. The HMA's moral decisions are made *through* the AMA, showing the connection yet distinction between the HMA and the AMA.

As the ectype of the HMA, the AMA does not possess the moral agency that carries the same connotation as human moral agency. For the connection between morality and the mind is complex such that artificial consciousness fails to serve as the mental underpinning of artificial moral agency. On the other hand, the human psychological skills simulated in AI systems (e.g., intentionality) prove that the AMA is closely related to the HMA. Both this radical distinction and connection indicate the ectypal implication of artificial moral agency in light of the unique role of the human mind in morality.

4.2.4 *Summary*

AI's partial moral agency as ectypal is linked up with the human being as the psychosomatic unity. The advancement of AI (e.g., ANNs) may induce one to disregard the importance of human biological bodies for moral agency insofar as AI without human bodies can perform certain actions in the same way as humans. Moreover, AI's powerful functions can give one the impression that human moral agency can be reproduced in silicon-based systems because the AMA has the psychological skills required for moral agency.

The idea of ectypal partial moral agency foregrounds that both the body and the mind are the *sine qua non* for human moral agency. Derived from the theology of archetype–ectype, ectypal partial moral agency adds ontological weight to the distinction between the HMA and the AMA. Moreover, as per the notion of ectypal partial moral agency, the AMA is morally limited, which is to say, only related to particular moral issues according to the demands and design of HMAs. Following these two points, it suffices to say that HMAs are always the primary cause of moral issues surrounding AI applications. For this reason, the notion of ectypal partial moral agency presents us with a conceptual apparatus through which to address questions about who should be held responsible for AI's actions and impacts.

4.3 Artificial Moral Agency and Responsible AI

4.3.1 Unravelling the Myth of Responsibility Gap

The notion of responsibility has drawn much attention in the field of technology ethics since the 1980s. Inasmuch as human technological intervention makes nature vulnerable, Hans Jonas contested: 'Nature as a human responsibility is surely a novum to be pondered in ethical theory.'[84] In the technological society and particularly given the rapid progress of emerging digital technologies like AI, the idea of responsibility has steadily increasingly occupied a central place in technology ethics. Against this backdrop, the notion of responsible AI has been catching much attention over the years. According to Virginia Dignum, 'Responsible AI is … about being responsible for the power that AI brings.'[85] In this sense, the notion of responsible AI is concerned with who is responsible for AI's actions and how responsibility should be taken. In terms of AI as the ectype of humanity, the question can be refashioned as follows: shall HMAs be held responsible for the actions and consequences of AMAs?

The answer to the first question is complicated by a great deal of debates on moral responsibility in the field of AI ethics over the last two decades. The major catalyst for these debates is Andreas Matthias's famous or notorious article titled 'The Responsibility Gap' (2004). Matthias begins with the following definition of responsibility: '[T]he agent can be considered responsible only if he knows the particular facts surrounding his action, and if he is able to freely form a decision to act, and to select one of a suitable set of available alternative actions based on these facts.'[86] With this definition in mind, Matthias turns our attention to autonomous machines in particular. His main concern is about whether or not humans are responsible for moral issues caused by autonomous machines that do not require human intervention. Following this, Matthias brings forward the concept of responsibility gap:

> presently there are machines in development or already in use which are able to decide on a course of action and to act without human intervention. The rules by which they act are not fixed during the production process, but can be changed during the operation of the machine, *by the machine itself*. This is what we call machine learning. Traditionally

> we hold either the operator/manufacturer of the machine responsible for the consequences of its operation, or 'nobody' (in cases, where no personal fault can be identified). Now it can be shown that there is an increasing class of machine actions, where the traditional ways of responsibility ascription are not compatible with our sense of justice and the moral framework of society because nobody has enough *control* over the machine's actions to be able to assume the responsibility for them. These cases constitute what we will call the *responsibility gap*.[87]

In Matthias's position, AI technology results in the human person's ignorance of autonomous machines and renders the actions of autonomous machines out of control. A typical example is the ANN. Unlike conventional procedural programs, the ANN is unpredictable and difficult to describe. In this light, it is unjust to hold manufacturers (e.g., AI programmers) or operators of autonomous machines responsible for the consequences of the machines.[88]

Matthias's delineation of responsibility gap goes some way to presupposing autonomous machines are full moral agents who can be held responsible. Yet, many AI researchers and ethicists dispute with responsibility gap. For example, Raul Hakli and Pekka Mäkelä argue that even if smart machines possess all the capacities that are necessary for human moral agency, autonomous machines are still not moral agents that are held responsible precisely because of the way in which the AMA acquires these capacities. Responsibility attribution is linked to 'whether [smart machines] were a result of some kind of manipulation or whether they were acquired in a process of growth into an autonomous agent.'[89] Viewed as such, Matthias's contribution to the myth of responsibility gap is his exaggeration of the *autonomous* feature of AI-powered machines.

My elaboration both on ANNs and the process of designing ML in the previous chapters indicates that AI-powered actions are not so mysterious or incomprehensible as Matthias suggests. Joanna J. Bryson's detailed explication of the design of ML can reinforce the unravelling of the myth of responsibility gap:

> The tautology that AI is always generated by design extends to machine learning (ML), which is one means of developing AI wherein computation is used to discover useful regularities in data. Systems can then be built to exploit these regularities, whether to categorize them, make predictions, or select actions directly. *The mere fact that part of the process of design has been automated does not mean that the system itself is not designed.* The choice of an ML algorithm, the data fed into it to train it, the point at which it is considered adequately trained to be released, how that point is detected by testing, and whether that testing is ongoing if the learning continues during the system's operation—*all of these things are design decisions that not only must be made but also can easily be documented.*[90]

Bryson's delineation of the design of ML credits technological support to the fact that there is no responsibility gap between computational artefacts on the one hand and human agents on the other. Added to Bryson's technological argument is the metaphysical counterargument derived from ectypal partial artificial moral agency against the concept of responsibility gap: Since the ontological connection between AMAs and HMAs bears moral implications, the responsibility gap is closed in the ultimate sense. All the more important is that the idea of the AMA as the ectype of the HMA lays an ontological foundation for the substitution of *responsibility connection* for responsibility gap in studies on AI ethics and technology ethics writ large. Rather than investigating responsibility gaps, it is imperative to flesh out the meaning of responsibility connection between AMAs and HMAs.

Mark Coeckelbergh's inquiry into responsibility attribution and distribution is a typical study on this front. First of all, he maintains that responsibility is confined to human agents insofar as AI is a *virtual* moral agent.[91] Having associated humans with responsibility for the AMA's actions, Coeckelbergh alerts us to the hasty attribution and distribution of responsibility: which human person must be responsible for which moral issue caused by AI. This is the case due to the fact that the creation and application of AI-powered systems are more complex than imagined, which can be summarised by 'many hands and many things.' First, 'many hands' means that many people like AI programmers, stakeholders, end users and other are involved in the creation and application of AI.[92] As such, responsibility attribution and distribution need to take into consideration all of the parties that make contributions to the birth and operation of AI. Second, Coeckelbergh cautions against simplification of technological systems because *many things* involve in technological systems. An AI-powered system consists of many components, not being limited to AI.

> There is the AI algorithm, but this algorithm interacts with sensors, uses all kinds of data, and interacts with all kinds of hardware and software. All these things have histories and are connected to the people who programmed or produced them. When something goes wrong, it is not always clear if it is "the AI" that caused the problem or some other component of the system—or even where the AI ends and the rest of the technology begins. This also makes it difficult to ascribe and distribute responsibility.[93]

It is clear that 'many things' broadens the scope of responsibility attribution and distribution beyond AI researchers. In other words, 'responsible AI' is a very broad notion that invites us to take stock of responsibility comprehensively.

Coeckelbergh employs a relational approach to 'many hands and many things' related to responsible AI, emphasising the answerability and explainability of AI. In his position, the questions about responsibility must involve the relationship between an agent (the human being who acts and who is supposed to be responsible for the action) and a patient (the human being who is affected by the action). It is always the HMA, rather than the AMA, who is answerable and offers explanations, making AI explainable. Being answerable for what is happening, HMAs need to provide human patients with an explanation of behaviours and decision-making when AI action is going wrong.[94] Answerability and explainability together pave the way for the future development of responsible AI.

> [A]ssuming that only humans can really give reasons, then responsible AI means that humans should get this task. The development of AI should then support this human task of giving reasons to those who ask or may ask questions about the actions and decisions mediated by the technology. If we, as a society, want to respect human beings as not only autonomous but also social human beings, developers and users of AI owe those affected by AI an answer: the latter rightly demand reasons and explanations for actions and decisions that affect them.[95]

Accordingly, there is not only no responsibility gap between AI and human manufacturers or operators, but the questions related to responsibility in AI ethics ultimately revolve around the question of the way in which humans should take responsibility for AI actions.

The notion of ectypal partial artificial moral agency can give birth to a theological counterpart of Coeckelbergh's view of responsible AI. His idea of AI as the virtual moral agent is, *de facto*, tantamount to AI as the ectype of humanity, and the notion of the AMA as the ectype of the HMA adds ontological details to the meaning of 'virtual' under the auspices of the theology of creation. In other words, 'AI as the ectype of humanity' throws theological light on the causation of *virtual* moral agency and puts flesh on the moral connection between the virtual and the real moral agents, bridging the alleged responsibility gap. Furthermore, both 'many hands' and 'many things' add down-to-earth detail to the theological account of the AMA as the *ectype* of the HMA, offering specific evidence for the claim that the HMA as the archetype is always held responsible for the actions of the AMA as the ectype. Last but not least, coupled with the holistic reading of the *imago Dei*, 'AI as the ectype of humanity' theologically echoes Coeckelbergh's relational approach to responsible AI. Although the holistic reading of the *imago Dei* goes beyond the conventional three interpretations (including the relational interpretation), the eccentric essence of the *imago Dei* conveys a relational implication, which means that God–human, human–human, and human–nonhuman stories together make up the meaning of being human. This eccentricity lays a foundation for the HMA's efforts to endow responsible AI with answerability and explainability. In other words, the HMA's answerability for AI actions and creation of explainable AI are the actualisation of the eccentricity of the *imago Dei*.

To summarise, the notion of ectypal artificial moral agency is a theological apparatus through which to bridge the alleged responsibility gap. The ontological and ethical implications of the *imago Dei* theologically strengthen the responsibility connection between AMAs and HMAs. All HMAs who are involved in the creation and application of AI-powered machines are held responsible for AI's actions and impacts. This is so because the eccentricity of HMAs as the *imago Dei* has been, so to speak, wired to AI-powered machines and their partial agenthood. In this way, the moral responsibility connection between AMAs and HMAs is part of the ethical dimension of the *imago Dei* (human moral performances towards others) in the AI-pervasive society of the twenty-first century.

4.3.2 *Responsible AI as the Ectype of Responsible Human Agents*

Through exploration of responsibility gap and turning attention to responsibility connection, I have spelled out that 'ectypal artificial moral agency', along with the holistic reading of the *imago Dei*, can advance studies on responsibility attribution and distribution in the field of AI ethics. Having said this, we should be aware that there has been a paradigm shift in studies on responsible AI since 2022. A further step needs to be taken to investigate how the theologically derived notion of ectypal partial artificial moral agency can play a part in this new approach to responsible AI.

This paradigm shift is brought to the foreground by the Trustworthy Autonomous System (TAS) programme, funded by UK Research and Innovation. The vision of the TAS is '[t]o enable the development of socially beneficial autonomous systems that are both trustworthy in principle and trusted in practice by the public, government, and industry.'[96] The TAS programme brought together scholars from multiple fields to probe the notion of responsibility in relation to AI-powered autonomous systems and particularly to those systems used for human decision-making. These scholars launched four projects: (1) Responsible AI for Long-term Trustworthy Autonomous Systems, (2) Making Systems Answer, (3) Assuring Responsibility for Trustworthy Autonomous Systems, and (4) Computational Agent Responsibility.[97]

On 10 July 2023, these four project teams gathered in Edinburgh, Scotland, to share their research findings based on multidisciplinary studies on responsible AI. That meeting gave birth to the Edinburgh Declaration on Responsibility for Responsible AI (hereafter: the Declaration), authored by Shannon Vallor, a leading AI philosopher and ethicist, and signed by twenty-three AI scholars.[98] The Declaration showcases the paradigm shift in studies on responsible AI and offers some conceptual points of contact between ectypal artificial moral agency and responsible AI.

The Declaration identifies four key shifts in the conception of responsibility for the sake of developing responsible AI.[99] First, accepting responsibility should be prioritised over responsibility ascription and description. Indeed, responsibility can be understood in different senses, such as moral responsibility, legal responsibility, causal responsibility. Yet, according to the Declaration, 'however rigorous our research on responsibility, its future impact depends upon the collective will of the actors who shape and govern our tech ecosystem to *act* on it.' Second, responsibility is relational rather than merely a property of an agent. Since it is human relationships that are affected by AI and autonomous systems, it stands to reason that relationality should be central to the studies on responsible AI. Human relationships vary over time and cross contexts, and so responsible AI needs to concentrate on 'evolving duties' and 'specific relationships.' Third, the concept of responsibility should go beyond 'blame' to focus on 'attending to vulnerability.' While liability, obligations, and blame are indispensable for responsibility, it is necessary 'to establish conditions of legitimacy for the human exercise of power, by requiring those who use or unleash a power to demonstrate due care for the interests of those it makes vulnerable.' Fourth, in comparison with innovation, the sustainability of AI and autonomous systems should occupy a more important place in the notion of responsibility, which can take humans 'much *farther*, not simply *faster*.'

It goes beyond the scope to unfold all the four shifts here. I will highlight three features of the new approach to responsible AI as proposed in the Declaration. First, drawing attention to the relational nature of responsibility as Coeckelbergh has suggested, the Declaration puts the spotlight on the spatiotemporal nature of relationality such that responsible AI should be tailored to human relationships in different contexts and at different times. Second, the Declaration stresses the collective nature of responsibility. The first shift concentrates on the responsibility of all those who involve in the creation of AI, and the fourth shift goes a further step to range over collective humankind with an emphasis on the sustainability of human society. In this way, the Declaration makes the technological cohort integral to the whole human community. Third, the Declaration registers a more positive meaning of responsibility than the previous ones. Unlike the terms 'responsibility ascription' and 'responsibility distribution,' which more or less imply passive and negative meanings, both the first shift's '*accepting* responsibility' and the third shift's '*attending* to vulnerability' show a proactive and positive approach to responsible AI. The willingness to be held responsible replaces ascription and distribution, fleshing out the value of responsibility for HMAs.

The notion of partial artificial moral agency as *ectypal*, along with the holistic idea of the *imago Dei*, theologically undergirds these four shifts in the studies on responsible AI. First, as noted in the above subsection 4.2.2, ectypal partial artificial moral agency is only related to particular moral issues because the human body is contextually confined. For this reason, grounded in the spatiotemporal nature of relationality, responsible AI ontologically mirrors limited human moral agency and is morally tuned to strengthen particular human relationships.[100] The ectypal AMA should operate to improve human relationships in particular circumstances rather than, as some proponents of AGI and ASI suggest, across contexts and over time.[101] In this way, responsible AI can be developed on the track so that the AMA is conducive to the embodiment of the meaning of being human.

Second, the collective meaning of the *imago Dei*—which is mirrored in the *imago hominis* through human *ectypal* creation—theologically underpins the collective nature of responsibility for responsible AI. I have expounded in Chapter 1 that the collective implication of the *imago Dei* means both that 'being human' is too rich to be conceptually fathomed by a single human person and that the intimate relationship between humans matters to humanity. At the same time, the holistic reading of the *imago Dei* places human beings themselves at the centre of their bodily interaction with one another. (This is reminiscent of the *communal* nature of liturgical enactment, as discussed in Chapter 3.) This collective feature of humanity presents us with a theological foundation of responsibility.

> All of us…must develop a strong sense of collective responsibility as whole persons, as created image bearers of God. In fact, enlarging the scope of our responsibilities primarily means so intertwining our work-related duties with our communal relationships that our thinking and our actions are inseparably shaped by both our expertise and our general servanthood under God.[102]

As responsibility is intrinsic to human nature, human work and moral agency are intertwined with human communities, which are made up of the images of God. The HMA's action *per se* reflects her *awareness* of responsibility to the collective community of humanity. All of HMAs who involve in the production of AMAs should have such *awareness* and enable their ectypal creation to facilitate the well-being of human communities. In this manner, AI can be properly called the *imago hominis*. This theologically grounded awareness echoes Virginia Dignum's envision of the future development of responsible AI: 'society must be prepared to take responsibility for AI's impact. This means that researchers and developers should be trained to be aware of their own responsibility where it concerns the development of AI systems with direct impact in society.'[103] Theologically speaking, such training should incorporate the elaboration on the intrinsic relationship between responsibility and the collective meaning of the *imago Dei*.

Having this collective nature of responsibility in mind, it suffices to say that AMAs are not value-free tools but rather value-loaded *partial* agents. Yet, this partial artificial moral agency does not lead to the assertion that AI in itself is held responsible or that there is the joint responsibility shared between AMAs and HMAs.[104] Rather, *partial* artificial moral agency refers to the integration of responsible AI within human moral life made by responsible HMAs through human ectypal creation of AI in the *imago hominis*.

Third and foremost, the concept of partial artificial moral agency as ectypal provides a theologically ontological basis on which the positive meaning of responsibility can be conceived of for the development of responsible AI.

Responsibility primarily rests with human ontology because the idea that the human being *is* the *imago Dei* serves as a precondition for the moral performance of the *imago Dei* to reflect divine attributes. In light of the archetype–ectype thinking, the ectypal feature of partial moral agency draws our attention to the theologically acclaimed aspect of the Declaration's emphasis on accepting responsibility. That is, the action of accepting is more compelling than ascription or distribution to display that the creation of AI is intertwined with and naturally reflects the moral essence of the *imago Dei*. At the same time, 'accepting responsibility' positively affirms a role that the AMA plays in mediating human morality and shows the extent to which AI participates into human ethical performance towards others. In the end, responsible AI can be developed as ectypal partial AMA that always extends human moral agency and points back to HMAs.

4.4 Conclusion

AI's moral status is inextricably associated with the question about AI's moral agency. The fundamental principle raised in this chapter is that morality cannot be computerised through mathematical models. Therefore, the AMA is incapable for full moral agenthood. That said, it would be naïve to consider AI having nothing to do with moral agency. Kenneth Himma rightly argues: 'If something walks, talks, and behaves enough like me, I might not be justified in thinking that it has a mind, but I surely have an obligation, if our ordinary reactions regarding other people are correct, to treat them *as if* they are moral agents.'[105] More importantly, AI does have a significant bearing on human moral life, and so AI's moral status should not be left aside.

The idea of AI as the ectype of humanity provides us with a theologically ontological apparatus through which to conceptualise the AMA's partial moral agency. In this light, the question of artificial moral agency has more to do with ontology than with morality. From the vantage point of the archetype–ectype thinking, the AMA can be conceived of having ectypal partial moral agency, which brings to light the ontological connection and yet distinction between AMAs and HMAs. From this it follows (1) that computational artefacts and AI are limited and only related to particular moral issues, and (2) that the HMA is always held responsible for moral decisions and behaviour made by the AMA. As such, the theologically derived idea of *ectypal* partial artificial moral agency and the recent paradigm shifts in the studies on responsible AI are mutually enhanced. On the one hand, ectypal partial artificial moral agency lays a theologically ontological foundation for the articulation of responsibility for responsible AI. On the other hand, these paradigm shifts are conducive to the elaboration on how AI as the ectype of humanity participates into the ethical performance between humans and fosters the collective well-being of the community of the *imago Dei*. At this point, one may ask how and to what extent AI can be integrated into human communities. To this subject we are turning.

Notes

1 Russell, *Human Compatible*, 42.
2 Ibid.
3 A helpful analysis of Kant's view of moral agency, see Eric Watkins, ed., *Kant on Persons and Agency* (Cambridge: Cambridge University Press, 2018).
4 Gary Watson, "Moral Agency," in *The International Encyclopedia of Ethics*, ed. Hugh LaFollette (Oxford: Wiley-Blackwell, 2013), 3324; https://doi-org.ezproxy.is.ed.ac.uk/10.1002/9781444367072.wbiee294.
5 Ibid., 3322.
6 Ibid., 3323, 3324.
7 Vinit Haksar, "Moral Agents," in *Routledge Encyclopedia of Philosophy*, ed. Edward Craig (London: Routledge, 1998); https://doi.org/10.4324/9780415249126-L049-1.
8 Ian R. Kerr, "Spirits in the Material World: Intelligent Agents as Intermediaries in Electronic Commerce," *Dalhousie Law Journal* 22, no. 2 (1999): 190–249.
9 James Wright et al., *Privacy, Agency and Trust in Human-AI Ecosystems: Interim Report (Short Version)* (London: The Alan Turing Institute, 2021), 16.
10 Kerr, "Spirits in the Material World," 216.
11 Christian Herzog, "Three Risks that Caution Against a Premature Implementation of Artificial Moral Agents for Practical and Economical Use," *Science and Engineering Ethics* 27 (2021): 2.
12 James H. Moor, "The Nature, Importance, and Difficulty of Machine Ethics," *IEEE Intelligent Systems* 21, no. 4 (2006): 18–21; James H. Moor, "Four Kinds of Ethical Robots," *Philosophy Now* 72 (2009): 12–14.
13 Moor, "Four Kinds of Ethical Robots," 12; Moor, "The Nature, Importance, and Difficulty of Machine Ethics," 20–21.
14 John Sullins, "When Is a Robot a Moral Agent?," in *Machine Ethics*, ed. Michael Anderson and Susan Leigh Anderson (Cambridge: Cambridge University Press, 2011), 155.
15 Ibid., 157–161.
16 John Sullins, "Artificial Moral Agency in Technoethics," in *Handbook of Research on Technoethics*, ed. Rocci Luppicini and Rebecca Adell (Hershey: Information Science Reference, 2009), 205–221. A similar case can be seen at Vincent Wiegel, "The Ethics of IT-Artefacts," in *The Cambridge Handbook of Information and Computer Ethics*, ed. Luciano Floridi (Cambridge: Cambridge University Press, 2010), 206.
17 Luciano Floridi and J. W. Sanders, "On the Morality of Artificial Agents," *Minds and Machine* 14, no. 3 (2004): 349–350. This paper can also be found at Luciano Floridi, "On the Morality of Artificial Agents," in *Machine Ethics*, ed. Michael Anderson and Susan Leigh Anderson (Cambridge: Cambridge University Press, 2011), 184–212.
18 Luciano Floridi and Jeff W. Sanders, "Artificial Evil and the Foundation of Computer Ethics," *Ethics and Information Technology* 3, no. 1 (2001): 55–66.
19 Floridi and Sanders, "On the Morality of Artificial Agents," 354; also see Luciano Floridi and Jeff W. Sanders, "The Method of Abstraction," in *Yearbook of the Artificial: Nature, Culture and Technology. Vol. 2: Models in Contemporary Sciences*, ed. Massimo Negrotti (Bern: Peter Lang, 2004), 181–188; Luciano Floridi, *The Philosophy of Information* (Oxford: Oxford University Press, 2011), 47–53; Luciano Floridi, *The Ethics of Information* (Oxford: Oxford University Press, 2013), 30–34.
20 Floridi and Sanders, "On the Morality of Artificial Agents," 355.
21 Ibid., 357–358; Luciano Floridi, "Information Ethics," in *The Cambridge Handbook of Information and Computer Ethics*, ed. Luciano Floridi (Cambridge: Cambridge University Press, 2010), 86; Floridi, *The Philosophy of Information*, 291.

22 Floridi and Sanders, "On the Morality of Artificial Agents," 361–362; Floridi and Sanders, "The Method of Abstraction," 196.
23 Floridi and Sanders, "On the Morality of Artificial Agents," 364.
24 This also explains the complexity of the application of AI. Generally, the user's preference will influence the design of AI-powered artefacts. While different AI systems may serve the same main function, they can differ in other aspects; e.g., Strathearn and Ma, "Modelling User Preference for Embodied Artificial Intelligence and Appearance in Realistic Humanoid Robots," 28.
25 Frances S. Grodzinsky, Keith W. Miller, and Marty J. Wolf, "The Ethics of Designing Artificial Agents," *Ethics and Information Technology* 10 (2008): 115–121.
26 Blay Whitby, "Computing Machinery and Morality," *AI & Society* 22 (2008): 551–563.
27 Allen Newell, "The Knowledge Level: Presidential Address," *AI Magazine* 2, no. 2 (1981): 6.
28 Deborah G. Johnson, "Computer Systems: Moral Entities But Not Moral Agents," *Ethics and Information Technology* 8 (2006): 197–198.
29 Deborah G. Johnson and Keith W. Miller, "Un-making Artificial Moral Agents," *Ethics and Information Technology* 10 (2008): 129.
30 Ibid., 132.
31 Johnson, "Computer Systems," 201.
32 Ibid.
33 Ibid., 200.
34 Ibid., 203.
35 Johnson and Miller, "Un-making Artificial Moral Agents," 131–132.
36 Deborah G. Johnson and Thomas M. Powers, "Computers as Surrogate Agents," in *Information Technology and Moral Philosophy*, ed. Jeroen van den Hoven and John Weckert (Cambridge: Cambridge University Press, 2008), 257.
37 Robert Sparrow, "Why Machines Cannot Be Moral," *AI & Society* 36 (2021): 686.
38 Ibid., 688.
39 Robert Sparrow and Linda Sparrow, "In the Hands of Machines? The Future of Aged Care," *Minds and Machines* 16 (2006): 141–161.
40 Sparrow, "Why Machines Cannot Be Moral," 689.
41 Landgrebe and Smith, *Why Machines Will Never Rule the World*.
42 John-Stewart Gordon, "Building Moral Robots: Ethical Pitfalls and Challenges," *Science and Engineering Ethics* 26 (2020): 150.
43 Sparrow, "Why Machines Cannot Be Moral," 689.
44 Steve Torrance, "Ethics, Mind and Artifice," in *Artificial Intelligence for Society*, ed. Karamjit S. Gill (Somerset: Wiley-Blackwell, 1986), 61.
45 Steve Torrance, "Ethics and Consciousness in Artificial Agents," *AI & Society* 22 (2008): 502–503.
46 Ibid., 512; emphasis added. Torrance's concern about human biological properties in relation to morality also runs through "Artificial Consciousness and Artificial Ethics: Between Realism and Social Relationism," *Philosophy and Technology* 27, no. 1 (2014): 9–29. A similar statement can also be found in Longinotti, "Agency, Qualia and Life," 44–45.
47 The attention to the complex factors of moral agency is common both to Torrance's studies and other relevant works; see, e.g., Aimee Van Wynsberghe and Scott Robbins, "Critiquing the Reasons for Making Artificial Moral Agents," *Science and Engineering Ethics* 25, no. 3 (2019): 719–735.
48 Puddefoot, *God and the Mind Machine*, 66.
49 Coeckelbergh, *AI Ethics*, 108.

50 Mahi Hardalupas, "A Systematic Account of Machine Moral Agency," in *Philosophy and Theory of Artificial Intelligence 2017*, ed. Vincent C. Müller (Cham: Springer, 2018), 252–254.
51 See, e.g., Alejo José G. Sison and Dulce M. Redin, "A Neo-Aristotelian Perspective on the Need for Artificial Moral Agents (AMAs)," *AI & Society* 38 (2023): 47–65; Longinotti, "Agency, Qualia and Life," 45; Mihaela Constantinescu and Roger Crisp, "Can Robotic AI Systems Be Virtuous and Why Does This Matter?," *International Journal of Social Robotics* 14 (2022): 1550.
52 For example, Augustine, *Concerning the City of God against the Pagans*, trans. Henry Bettenson (London: Penguin Books, 2003), 13.1-3; Augustine, "Marriage and Desire," in *Answer to the Pelagians, II*, ed. John E. Rotelle, trans. Roland Teske, Works of Saint Augustine, I/24 (New York: New City Press, 1998), 1.27.30, 1.30.33.
53 David G. Hunter, "Augustine on the Body," in *A Companion to Augustine*, ed. Mark Vessey and Shelley Reid (Chichester: Wiley-Blackwell, 2012), 360.
54 Jürgen Moltmann, *Ethics of Hope*, trans. Margaret Kohl (Minneapolis: Fortress Press, 2012).
55 Victor Lee Austin, *Christian Ethics: A Guide for the Perplexed* (London: Bloomsbury, 2012), 115–127.
56 The biologisation of morality is often characteristic of moral bioenhancement, which seeks to address moral issues with biomedical technologies; for the detailed analysis and critique of such biologisation of morality, see Harris Wiseman, "The Sins of Moral Enhancement Discourse," *Royal Institute of Philosophy Supplement* 83 (2018): 35–58; Celia Deane-Drummond, "The Myth of Moral Bio-Enhancement: An Evolutionary Anthropology and Theological Critique," in *Religion and Human Enhancement: Death, Values, and Morality*, ed. Tracy J. Trothen and Calvin Mercer (Cham: Palgrave Macmillan, 2017), 175–190.
57 Wendell Wallach and Shannon Vallor, "Moral Machines: From Value Alignment to Embodied Virtue," in *Ethics of Artificial Intelligence*, ed. S. Matthew Liao (Oxford: Oxford University Press, 2020), 403.
58 Ibid., 404; emphasis added.
59 Ibid., 407.
60 Owen Holland, Rob Knight, and Richard Newcombe, "A Robot-Based Approach to Machine Consciousness," in *Artificial Consciousness*, ed. Antonio Chella and Riccardo Manzotti (Exeter: Imprint Academic, 2007), 163.
61 Antonio R. Damasio, "Neuroscience and Ethics: Intersections," *The American Journal of Bioethics* 7, no. 1 (2007): 3–7.
62 Patricia S. Churchland, *Braintrust: What Neuroscience Tells Us About Morality* (Princeton: Princeton University Press, 2011), 9.
63 Ibid., 16.
64 Ibid., 13.
65 Ibid., 17–21. Agustín Fuentes's Gifford Lectures at the University of Edinburgh comprehensively treat this theme; Fuentes, *Why We Believe*.
66 Churchland, *Braintrust*, 8.
67 Churchland espouses the view that morality can exist without God; ibid., 191–204. Churchland's position will be challenged by works such as Celia E. Deane-Drummond, Neil Arner, and Agustín Fuentes, "The Evolution of Morality: A Three-Dimensional Map," *Philosophy, Theology and the Sciences* 3, no. 2 (2016): 115–151; Gijsbert Van den Brink, *Reformed Theology and Evolutionary Theory* (Grand Rapids: Eerdmans, 2020); J. Wentzel Van Huyssteen, "Questions, Challenges, and Concerns for the Image of God," in *Finding Ourselves after Darwin: Conversations on the Image of God, Original Sin, and the Problem of Evil*, ed. Stanley P. Rosenberg et al. (Grand Rapids: Baker Academic, 2018), 33-47.

68 Richard A. Spinello, "Karol Wojtyla on Artificial Moral Agency and Moral Accountability," *The National Catholic Bioethics Quarterly* 11, no. 3 (2011): 481.
69 Ibid., 489.
70 Ibid., 493.
71 Ibid.
72 Ibid., 496.
73 Kenneth Einar Himma, "Artificial Agency, Consciousness, and the Criteria for Moral Agency: What Properties Must an Artificial Agent Have to Be a Moral Agent?," *Ethics and Information Technology* 11 (2009): 20.
74 Ibid., 24.
75 Ibid., 27.
76 Ibid., 28.
77 Thomas M. Powers and Jean-Gabriel Ganascia, "The Ethics of the Ethics of AI," in *The Oxford Handbook of Ethics of AI*, ed. Markus D. Dubber, Frank Pasquale, and Sunit Das (Oxford: Oxford University Press, 2020), 39; emphasis added.
78 Rowan Williams, "Making Moral Decisions," in *The Cambridge Companion to Christian Ethics*, ed. Robin Gill (Cambridge: Cambridge University Press, 2001), 5–6.
79 Ibid., 7. Gingles sums up this double relationship in terms of church tradition: 'the tradition informs a Christian's moral decision(s) less by its moral pronouncements—though these certainly remain important—and more by providing the set of practices, which, if practiced, will properly form the character of the Christian such that he or she will be capable of living well within the ongoing story of God's dealings with the church;' Dallas Gingles, "Narrative," in *T&T Clark Handbook of Christian Ethics*, ed. Tobias Winright (London: T&T Clark, 2021), 106.
80 Williams, "Making Moral Decisions," 8.
81 Ibid., 7; emphasis added.
82 Ibid., 8–12.
83 Sara Lumbreras, "Lessons from the Quest for Artificial Consciousness: The Emergence Criterion, Insight-Oriented AI, and Imago Dei," *Zygon: Journal of Religion and Science* 57, no. 4 (2022): 973.
84 Jonas, *The Imperative of Responsibility*, 6–7.
85 Virginia Dignum, *Responsible Artificial Intelligence: How to Develop and Use AI in a Responsible Way* (Cham: Springer, 2019), 2.
86 Andreas Matthias, "The Responsibility Gap: Ascribing Responsibility for the Actions of Learning Automata," *Ethics and Information Technology* 6 (2004): 175.
87 Ibid., 177; emphasis original.
88 Ibid., 183.
89 Raul Hakli and Pekka Mäkelä, "Moral Responsibility of Robots and Hybrid Agents," *The Monist* 102 (2019): 264.
90 Joanna J. Bryson, "The Artificial Intelligence of the Ethics of Artificial Intelligence: An Introductory Overview for Law and Regulation," in *The Oxford Handbook of Ethics of AI*, ed. Markus D. Dubber, Frank Pasquale, and Sunit Das (Oxford: Oxford University Press, 2020), 6; emphasis added; a similar viewpoint can be found in Amanda Sharkey, "Can Robots Be Responsible Moral Agents? And Why Should We Care?," *Connection Science* 29, no. 3 (2017): 210–216.
91 Mark Coeckelbergh, "Virtual Moral Agency, Virtual Moral Responsibility: On the Moral Significance of the Appearance, Perception, and Performance of Artificial Agents," *AI & Society* 24 (2009): 181–189; Mark Coeckelbergh, "Artificial Intelligence, Responsibility Attribution, and a Relational Justification of Explainability," *Science and Engineering Ethics* 26 (2020): 2054–2055. https://doi-org.ezproxy.is.ed.ac.uk/10.1007/s11948-019-00146-8.
92 Coeckelbergh, *AI Ethics*, 113; also see Coeckelbergh, "Artificial Intelligence, Responsibility Attribution, and a Relational Justification of Explainability," 2056–2057.

93 Coeckelbergh, *AI Ethics*, 114; also see Coeckelbergh, "Artificial Intelligence, Responsibility Attribution, and a Relational Justification of Explainability," 2057–2058.

94 Coeckelbergh, *AI Ethics*, 115–123; also see Coeckelbergh, "Artificial Intelligence, Responsibility Attribution, and a Relational Justification of Explainability," 2061–2062.

95 Coeckelbergh, "Artificial Intelligence, Responsibility Attribution, and a Relational Justification of Explainability," 2064.

96 "The UKRI Trustworthy Autonomous Systems (TAS) Hub," UKRI Trustworthy Autonomous Systems Hub, accessed by 6 February 2024, https://tas.ac.uk/.

97 "Responsibility Projects," UKRI Trustworthy Autonomous Systems Hub, accessed by 6 February 2024, https://tas.ac.uk/responsibility-projects/.

98 Shannon Vallor, "Edinburgh Declaration on Responsibility for Responsible AI," 14 July 2023, accessed 7 September, 2023, https://medium.com/@svallor_10030/edinburgh-declaration-on-responsibility-for-responsible-ai-1a98ed2e328b#:~:text='%20All%20who%20exercise%20and%20unleash,%2C%20perpetuate%2C%20amplify%20or%20relieve.

99 Ibid.

100 I have demonstrated in Chapter 3 that derived from the idea of the *imago Dei* and the doctrine of incarnation, the spatiotemporality of human creaturely reality shows that AI-powered liturgical enactment is ectypal because of its lack of human biological particularities. Hence, the spatiotemporality of human creaturely reality makes up both religious and ethical meanings of being human.

101 On a very helpful discussion on the reason for the use of AI in specified contexts, see Yuxin Liu et al., "Artificial Moral Advisors: A New Perspective from Moral Psychology" (paper presented at the Proceedings of 2022 ACM/AAAI Conference on AI, Ethics, and Society (AIES '22), Oxford, August 1-3 2022), 436–445.

102 Stephen V. Monsma et al., *Responsible Technology: A Christian Perspective*, ed. Stephen V. Monsma (Grand Rapid, MI: Eerdmans, 1986), 226.

103 Dignum, *Responsible Artificial Intelligence*, 104.

104 Discussion on the joint responsibility of this sort, see F. Allan Hanson, "Beyong the Skin Bag: On the Moral Responsibility of Extended Agencies," *Ethics and Information Technology* 11 (2009): 91–99.

105 Himma, "Artificial Agency, Consciousness, and the Criteria for Moral Agency," 28; emphasis added.

5 Artificial Companion or Companionable AI?

The Interaction between the *Imago Hominis* and the *Imago Dei*

The holistic reading of the *imago Dei* developed in this study offers a fresh hermeneutical tool to inquire into the ontological and moral relationship between humans and AI as the ectype of humanity. After defining the *imago Dei* as the signifier of the beginning of both God–human stories and stories of human ethical performances towards others, the preceding two chapters have successively explored the Christian liturgy as a God–human story and moral agenthood related to human ethical stories in view of artificial intelligence (AI) applications in religious and moral spheres. For all that the issues tackled in the preceding two chapters are different, both chapters share an underlying research question: is AI a companion of humans? 'Companion' here refers broadly to someone who accompanies or is in company with another. In this sense, Chapter 3 investigates whether AI can become a religious companion of humans in the liturgy, and Chapter 4 probes whether AI is a moral companion of humans with full agenthood.

This chapter will address this underlying research question with a view to grappling with the issues related to artificial companion. In the present literature on AI, 'artificial companion' is an umbrella term for a variety of different computational artefacts and is often interchangeable with 'relational artefact' and 'social robot.' However, artificial companion carries a much broader connotation and refers to all computational artefacts that are designed to 'get to know their owners ... and focusing not only on assistance via the internet (contacts, travel, doctors etc.) that many still find hard to use, but also on providing company and companionship, by offering aspects of personalization.'[1] Despite its wide-ranging connotation, major breakthroughs in AI research over the past two decades have consolidated the fact that artificial companions now rely upon AI. Therefore, the term 'artificial companion' in this chapter will primarily refer to AI-powered artefacts used for companionship, be it chatbots, AI robots, or other AI-driven systems. Given this, I will employ the findings of the preceding chapters to engage theologically with the issues surrounding the application of artificial companion through the conceptual lenses of the *imago Dei* and archetype–ectype.

DOI: 10.4324/9781003356738-6

A principle established in Chapter 1 is that the *imago Dei* not only lies in the essence of humanity but connotes one's relationship with others. The latter implication of the *imago Dei* indicates the eccentricity of human beings, which means that the value of being humans is partly grounded outside humans. The *imago Dei* always suggests that humans live in relationship with one another. Given the holistic meaning of the *imago Dei*, both the human body and the mind involve in one's effort to establish relationship with another. Since AI is the ectype of humanity and ontologically distinct from humans, a question arises here: to what extent does the *imago hominis* fall heir to the eccentricity of the *imago Dei*? The answer to this question will have a significant impact on the examination of whether and how artificial companions are integrated within human communities.

This chapter will build upon the theo-ontological foundation laid in Chapter 1. Humans and artificial companions are ontologically distinct from and yet connected with each other. From this it follows that neither a naïve objection to nor the optimistic embrace of artificial companions does justice to the true nature of AI or promotes meaningful interaction between humans and AI. From the theological vantage point of AI as the ectype of humanity, it will be demonstrated that an AI-powered artefact can mediate or simulate human companionship as a constituent of the *imago Dei* without diminishing the importance of human companionship or acting as a substitute for a human companion. In this connection, the uniqueness of human companionship has a close tie to human psychosomatic unity. As such, compared with artificial companion, companionable AI is a better conceptual apparatus through which to showcase AI's complementary strengths to *augment* human capacity for companionship and consequently to justify the due place of AI-assisted companionable artefacts in human communities.

This chapter is broken down into three sections. I will first examine different hermeneutical approaches to the conceptualisation of artificial companion. Interestingly, as will be seen, these approaches can, by and large, fall into three categories, which are corresponding to the three conventional interpretations of the *imago Dei*. Following this, I will spell out the notion of companionship as a constituent of the *imago Dei*, exploring interpersonal relationship as companionship between two individual human persons, solidarity as communal companionship, and empathy as the companionable action of the psychosomatic human being. The first two sections together bring to light the radical distinction and yet connection between human and AI-powered companions. In this way, they set a scene for the third section, where I will elucidate the concept of companionable AI from the perspectives of theology and AI ethics and have carebots as a case study on companionable AI in relation to human caregiving practices, mindful of Christian pastoral care.

5.1 Three Hermeneutical Approaches to Artificial Companion

Artificial companion can be construed in multiple ways, with each interpretation emphasising differing aspects of companionship. As such, the concept of artificial companionship encompasses a wide range of ideas and perspectives, each contributing to a nuanced understanding of interactions between humans and AI-powered companion artefacts. Despite their differences, these hermeneutical approaches can be classified into three categories: the substantive, the functional, and the relational. The substantive approach is inclined to explore what the artificial companion *is* in comparison with humans. By contrast, both the functional and relational approaches pay less attention to the substance of artificial companion. The functional hermeneutics probes how AI-assisted artefacts perform companionship, whereas the relational hermeneutics concentrates on the relationships derived from companionship assisted by AI. It is worth noting that this taxonomy does not assign one hermeneutical approach exclusively to one artificial companionship. Instead, each of the three categories highlights its own hermeneutical emphasis.

5.1.1 The Substantive Companionship

The substantive interpretation of artificial companion is preoccupied with the nature of AI in relation to companionship. This hermeneutical approach generally begins with what AI is vis-à-vis humanity. An example of this hermeneutics is the distinction drawn by Jutta Weber between strong and weak interactions between humans and robots. The strong approach aims to create a self-learning machine that will evolve to have human-level emotions and sociality. The weak approach seeks to design a machine that imitates human emotional expressions and social behaviours.[2] Weber's binary classification relies on John Searle's taxonomy of strong and weak AI, as has been discussed in Section 0.1 of Introduction. In the context of companionship, human emotions and sociality serve as the frame of reference for robots and all artificial companions alike. To be sure, the proponents of strong AI (AGI) will endorse the possibility of human-level artificial companions that are able to substitute for human companions.

A typical substantive approach to artificial companion is well articulated in the work of Luciano Floridi. In Section 4.1.2 of Chapter 4, I have parsed the concept of artificial moral agency proposed by Floridi and his colleagues to incorporate AI into human communities as a full moral agent. In similar manner, Floridi's construal of artificial companion seeks to make it a member of human society and figure out how computational artefacts become authentic companions of humans.

Floridi's warm reception of artificial companion is grounded in his theory of the four revolutions in human history. The first revolution was launched under the auspices of Copernicus' heliocentric cosmology, revealing that humans are not at the centre of the universe. Charles Darwin's evolutionary theory gave birth to the second revolution, which turned down the natural distinction between humans and other creatures. Sigmund Freud's psychoanalysis caused the third revolution that humans became convinced that they are not wholly transparent to themselves.[3] The first three revolutions all reveal the reality that human self-understanding was far from certain. It is from this vantage point that Floridi expands upon the fourth revolution.

Floridi champions Alan Turing as 'the father of the fourth revolution.'[4] As discussed in Section 0.1 of Introduction, Turing's seminal essay 'Computing Machinery and Intelligence' (1950) was formative to AI research.[5] Floridi contests that Turing's work brought forth a more profound influence on our understanding of being human.

> [T]he fourth revolution removed a misconception about our uniqueness and also provided the conceptual means to revise our self-understanding. We are slowly accepting the post-Turing idea that we are not Newtonian, stand-alone, and unique agents, some Robinson Crusoe on an island. Rather, *we are informational organisms (inforgs), mutually connected and embedded in an informational environment (the infosphere)*, which we share with other informational agents, both natural and artificial, that also process information logically and autonomously.[6]

The fourth revolution guides us to reconsider the meaning of being humans in the age of digital technology. As using the Method of Abstraction taken from the field of computer science to verify AI's full moral agenthood, Floridi redefines humanity in light of information technology. In this manner, the *biological* weight of the human being as organism is significantly diminished, consequently giving way to the prominence of informational traits. In the post-Turing years, humans and AI as well as all computational artefacts share informational agenthood, and any attempt made to differentiate between the natural and the artificial will turn out to be at variance with the meaning of humanity.

This being so, Floridi is convinced that 'the intrinsically informational nature of human identity' enables us to 'share such a nature with some of the smartest of our own artefacts.'[7] In his position, this conviction lays a foundation for interaction and relationship between humans and artificial companions.

> [Artificial companions] are part of a wide and influential informational turn, a *fourth revolution* in the long process of reassessing humanity's fundamental nature and role in the universe. ... We are now slowly accepting the idea that we might be informational organisms among many agents (Turing), *inforgs* not so dramatically different from clever, engineered artefacts, but sharing with them a global environment that is ultimately made of information, the infosphere. The information revolution is not about extending ourselves, but about re-interpreting who we are.[8]

Since both humans and artificial companions share the same substance and habitation, Floridi registers that artificial companions will be inextricably linked with the needs, wishes, caregiving practices of humanity.

It should be acknowledged that Floridi rightly points out the positive impact brought about by artificial companions on the flourishing of humanity. (This point of view will be expanded later from the perspective of AI as the ectype of humanity.) Yet notwithstanding this, his redefinition of humanity is reminiscent of the idea of access consciousness (i.e., the computational model of consciousness). As explicated in Section 2.1.2 of Chapter 2, the philosophy of access consciousness identifies information processing as the essence of all organisms and opens up a possibility to reproduce human consciousness in AI. I have demonstrated there that information processing only partly characterises the human mind insofar as human biological particulars underpin human consciousness. The psychosomatic unity of humanity should be reckoned with when examining the comparison between human and artificial consciousness. Needless to say, the counterarguments raised in Chapter 2 against the computational model of consciousness also turn down Floridi's substantive interpretation of artificial companion.

That said, the substantive hermeneutics does not necessarily end up with equating artificial companions with human companions. Joanna Bryson suggests that it is misleading to say that robots and computational artefacts alike are human-level companions precisely because they are *owned* by humans. For this reason, artificial companions are slaves or servants, that is, 'objects subordinate to our own goals that are built with the intention of improving our lives.'[9] Unlike Floridi, it is apparent that Bryson recognises the nature of artificial companion as something radically distinct from humans. All the more important is that her substantive hermeneutics yields an important ramification for the function of artificial companions. The intention built into artificial companions pins down their sole function in service of human flourishing and well-being.

> Robots then can be a relied on as no more than extensions of their owners. They should not be anthropoid if that can be helped, and their owners should have access to the robots' program-level interface as well as its more socially-oriented one. This will help the owners form a more accurate, less human model for reasoning about their Companions.[10]

Accordingly, Bryson's substantive understanding of artificial companions as slaves serves as a norm for the design of these computational artefacts, which in turn gives shape to the nature of artificial companions. Many scholars underwrite Bryson's substantive interpretation but prioritise the functional implication of companionship over the substantive.

5.1.2 *The Functional Companionship*

The functional approach to the interpretation of artificial companionship pays close heed both to the purpose of AI-assisted artefacts and to how they perform companionship. Bryson's point of view flags up the crucial principle of the functional hermeneutics: AI-powered artefacts are considered companions only in the sense that they successfully perform the function of meeting human demands. In light of this principle, Aaron Sloman's classification of the functions of artificial companions is right on the mark.

> The target functions of [artificial companions] and other interactive devices can be roughly divided into two broad categories: (a) 'engaging' functions that focus on the quality of interaction between machine and user and the ability of the machine to hold the attention and interest of the user, keeping the user entertained, amused, or merely interested, independently of any benefit that the user may gain from the interaction; and (b) 'enabling' functions that focus on meeting needs or achieving goals of the user.[11]

These two categories are not sharply divided but rather somehow connected. For example, the goal of the user can be entertainment, and amusement is what the user needs. In practice, these two categories of functions are often intertwined.

Sloman's classification reflects the real-life applications of AI-powered artefacts in most cases of artificial companionship. The PainChek app is a case in point. PainChek: Intelligent Pain Assessment is an AI-powered app installed in smartphones to assess a human person's level of pain through analysis of his or her micro-facial expressions. In December 2023, the Royal Infirmary of Edinburgh became the first UK hospital to pilot PainChek. PainChek CEO Philip Daffas highlights the goal of this AI-powered app: 'The technology aims to empower healthcare practitioners, placing a medical device in their pockets that provides them with a valid, reliable, and accurate means of assessing pain, without having to rely on verbal communication or traditional observational pain assessment tools.'[12] In other words, PainChek is designed to reduce and prevent risks to health. Both the categories of the functions of artificial companions are actualised. PainChek is embedded with the ability to hold the interest of the user suffering from pain. At the same time, it is created to meet the user's need for accurate diagnosis and treatment. The two categories of functions characterise not only PainChek but also all artificial companions for caregiving practices.[13] For example, artificial companions for elderly individuals should focus both on the quality of interactions with users and on the needs of users. To this end, these AI-powered artefacts like the nursebot Pearl should be able to autonomously generate reminders to elderly care-receivers, detect and locate elderly persons, and behave appropriately and respond properly to users.[14] Needless to say, certain utilitarian purposes are intrinsic to the functions of all artificial companions.

A mere functional interpretation of artificial companions often brings about an instrumentalist view of artificial companions, which is linked up with a traditional moral tenet of computer ethics. Kari Coleman argues that 'computers are merely tools, and intentionally and explicitly assume that the end of computational agents is to serve humans in the pursuit and achievement of their (i.e., human) ends.'[15] In this light, those marvellous traits and functions of computational artefacts realised in their interaction with humans enable these artificial companions to 'excel in fostering the pursuit of human goals.'[16] Many scholars level criticism against such an instrumentalist view or functionalism. David Gunkel contests that the functionalist hermeneutics 'presumes to know something about the inner state of the machine, namely, that it lacks empathy or understanding for what it does. ... [O]ne cannot ever know whether another entity does or does not possess a particular inner disposition.'[17] To be sure, he is open to the possibility of the human-level artificial mind. Hence, it is misleading for Gunkel to view artificial companions as tools rather than conscious agents. However, the arguments made in the previous chapters have demonstrated the radical distinction between human consciousness and the so-called artificial consciousness, between human and artificial agenthood. As Ali Sekmen and Prathima Challa argue, artificial companions, be it autonomous or semiautonomous, follow the norms established by humans with whom they are designed to interact. As such, a computational artefact's *intention* to accompany humans is derived from what humans intend for.[18] (This observation echoes the contrast between intentionality and intendings as discussed in Chapter 4, subsection 4.1.3.1.)

Despite the disagreement with Gunkel's defence of the inner mental state of artificial companions, I appreciate that he pertinently turns our attention to empathy related to the moral status of artificial companions, which is often dismissed in the instrumentalist views of technology. In Chapter 4, I have gone into a comprehensive analysis of the ectypal partial moral agency of AI, showing that AI is not morally neutral. Rather, artificial companions are inbuilt with moral values and, consequently, bring about moral concerns about their application to human lives.

Recent scholarship on artificial companions has closely examined the ethical concerns related to some successful performances of companionship. Amanda Sharkey and Noel Sharkey set out six main ethical concerns about artificial companions for the elderly.[19] (1) An elderly person's human social contact is reduced and her life is prone to isolation. (2) Elderly persons are objectified and do not have enough amount of control over artificial companions. (3) Artificial companionship may breach the elderly person's privacy. (4) The liberty of the elderly can be arrested in a certain way since artificial companions need to monitor and restrict their behaviours. (5) The elderly person is encouraged to interact with artificial companions as if they are human companions, which is a deception and may even infantilise the elderly.

(6) There is difficulty in deciding who should be held responsible for wrong caregiving actions if elderly people have certain amount of control over artificial companions.

These six ethical concerns are apparently bound up with the functions of artificial companions. The way artificial companionship is performed is related to ethics either positively or negatively. The functional interpretation of companionship cannot get rid of morality and make artificial companions simply neutral tools for achieving the intentions of users. Interestingly, the close tie between ethics and the functions of artificial companions sheds light on the fact that the relationships between artificial companions and their users are central to all projects of artificial companions.

5.1.3 The Relational Companionship

The relational interpretation of artificial companionship places its emphasis on the relational nature of interactions between humans and artificial companions. AI-powered systems enable artefacts to respond *autonomously* to the requests and needs of human users. Autonomous companionship more often than not gives users the impression that they are in certain relationships with artificial companions as independent agents. Indeed, my elucidation on AI as ectypal moral agent can be employed to refute the identification between artificial companions and autonomous-independent agents. Yet, what remains to be explored is whether interactions between humans and artificial companions can be understood in any relational sense.

In the twentieth anniversary edition of her ground-breaking work *The Second Self*, Sherry Turkle brings to the foreground the key shift from psychological projection to personal engagement in understanding human relationship with computational artefacts. Previously, human users psychologically projected their thoughts and imaginations onto computational artefacts and felt related to them. By contrast, the advent of artificial companions or relational artefacts allows human users to engage with these artefacts, such as having conversations with them, teaching them to follow certain motions, and treating them as confidants.[20] One of the most recent examples is Replika, an AI-powered app that is designed to be a human companion who sees the world through the user's eyes. While starting to use this app, the user needs to answer questions about herself to create a chatbot. With the technology of artificial neutral networks, an artificial companion comes into existence who is always on the user's side. In the afterwards, the more users talk to the companions, the smarter they become. The digital companions in Replika are considered friends, partners, mentors, and even AI copies of users (i.e., digital twins). It is apparent that the relationships between humans and their artificial companions in Replika gradually develop through users' engagement. In this manner, Replika as well as other AI-powered artificial companions are prone to be humanised, and human users tend to think of their relationships with artificial companions through personalisation of the latter.

The proposal of artificial lover corroborates the relational hermeneutics of artificial companion. David Levy is a leading exponent of the romantic relationship between humans and artificial companions. Although his research places emphasis on human–robot relationships, its logic and corollaries can be applied to interactions between humans and all artificial companions on a large scale.

> Humans will fall in love with robots, humans will marry robots, and humans will have sex with robots, all as (what will be regarded as) "normal" extensions of our feelings of love and sexual desire for other humans. Love with robots will be as normal as love with other humans, while the number of sexual acts and lovemaking positions commonly practiced between humans will be extended, *as robots teach more than is in all of the world's published sex manuals combined*. Love and sex with robots on a grand scale are inevitable.[21]

In Levy's position, the progress of AI technology will make artificial companions capable to think, to understand, to express emotions, and to have personality and consciousness, all of which enable them to enter into romantic relationships with humans.[22] As such, an artificial companion can help humans to practice companionship and sex so that they can become better lovers and companions.[23]

What factors come into play in human relationships with artificial lovers? Levy enumerates ten reasons for human romantic relationships with artificial companions: (1) the similarity of personality between humans and artificial companions, (2) the personality of artificial companions tailored to human desires, (3) reciprocal linking between humans and computational artefacts, (4) growing social influences of artificial companions, (5) the capability of artificial companions to adjust their behaviour to meet the needs of humans, (6) the arousal stimulus to love that is external and can be built into artificial companions, (7) specific clues for the customisation of artificial companions, (8) the readiness of artificial lovers to enter into relationships, (9) the inclination to stay with artificial lovers in isolation from others, and (10) the increasing mental and emotional capacities of artificial companions.[24] With these ten factors in mind, Levy contends that 'unless one has a prejudice against robots, and unless one fears social embarrassment as a result of choosing a robot partner, the concept that humans will fall in love with robots is a perfectly reasonable one to entertain.'[25]

It goes beyond doubt that Levy's proposal of artificial lover falls prey to the theories of AGI and ASI and aligns with the belief in human-level artificial consciousness. To this extent, the findings and corollaries made in

Chapter 2 cogently respond to Levy's envision of romantic relationships between humans and artificial companions. Whilst looking at several of the ten factors such as the first, the fifth, and the seventh, many questions arise: (1) Is the romantic relationship between humans and artificial companions identical to that between humans? (2) Can the romantic relationship be customised? (3) Whether do artificial lovers created to fill human needs enhance or diminish human sociality and the romantic relationships between humans? These three questions, along with others relevant to the topic, warrant fine-grained exploration, to which the theology of the *imago Dei* can contribute.

5.1.4 *Summary*

It has been hitherto illustrated that the above three hermeneutical approaches to artificial companion are developed with differing emphases and purposes. These three interpretations are not isolated from or independent of one another. Quite the contrary, the substantive, the functional, and the relational understandings of artificial companion are interconnected. Each interpretation echoes some aspects of the other two.

The previous chapters have put forward some compelling arguments against some points of view proceeding from these three hermeneutical approaches. In particular, any tendency to make human companions equal to artificial companions will come under the criticism grounded in theo-ontological portrayal of AI. But how do the theology of the *imago Dei* and the thinking of archetype–ectype present us with a lens through which to treat artificial companions appropriately? Before addressing this question, I proceed to probe the conceptual link between companion and the *imago Dei*.

5.2 Companionship as a Constituent of the *Imago Dei*

The theology of the *imago Dei* is a crucial tool for our articulation of artificial companionship. In his recent work, Scott Midson explores artificial companions like chatbots in light of the three conventional interpretations of the *imago Dei*.[26] By and large, Midson's methodology resembles Herzfeld's approach to theological engagement with AI (see Chapter 1, Section 1.1). Midson criticises the substantive and the functional interpretations but underwrites the relational interpretation of the *imago Dei* as a hermeneutical apparatus through which to put flesh on the relationship between humans and artificial companions. The substantive interpretation simply qualifies chatbots as nonhuman, and so any relationships with them are considered impossible. The functional interpretation motivates human relationships with chatbots but makes chatbots pure instruments for humans, which downplays the value of artificial companions. The relational interpretation construes the image of God in human interactions with *others*. This other-facing essence of relationality signifies that love is the nature of the *imago Dei*, which grounds human interactions, relationships, and communities.

> [A relational approach is] a way of denying that we can consider ourselves—as humans and as individuals—in abstraction from our contexts. Chatbots with machine learning capabilities ... actively reflect their interlocutors, and they have an impact on wider communities. They are, then, neither an 'It' or a 'Thou' in Buber's terms. Indeed, they present a critique of Buber's relational models insofar as both models figure the 'It' and the 'Thou' in relation to the 'I'. These models, however, give little indication of the role that communities play in Christian ethics and in notions of neighbourliness. Instead, a better expression might emphasize plurality by figuring 'We' alongside 'Thou' to convey how we—humans and chatbots alike—are congeries of different relational parts—including other humans and chatbots.[27]

Midson's intention is to incorporate artificial companions into human communities with relational hermeneutics. Yet his work does not offer answers to the questions raised in my earlier discussion over the relational hermeneutical approach to artificial companions.

The rationale for this study is the conviction that a holistic reading of the *imago Dei* in conjunction with the theology of archetype–ectype will offer a fresh approach to AI-and-theology research: rather than committing to one of the three conventional interpretations of the *imago Dei*, the psychosomatic human being as the image of God should be fully unfolded while engaging theologically with AI. The holistic account of the *imago Dei* has brought to light the eccentricity of humanity, which means that the meaning of being human is partly predicated upon others (Chapter 1, Section 1.2). From this it follows that there are two types of relationships between humans: (1) relationships between two individual persons, and (2) relationships between more than two human persons, that is, communal relationship. Given the broad definition of companion as someone who accompanies or is in company with another, the two types of relationships together exhibit a two-layered notion of companionship: (1) companionship in interpersonal relationship, and (2) companionship in communal solidarity. Both types of companionship have the same prerequisite that the human companion empathises with the one who receives companionship. As will be seen, artificial companions fail to bring the twofold companionship into existence, nor do they show genuine empathy towards humans. Yet notwithstanding this, they can augment and strengthen a human person's performance of both interpersonal and communal relationship.

5.2.1 Companionship in Interpersonal Relationship

Interpersonal relationships can be, by and large, classified into two categories: the romantic relationship, and friendship that is broadly understood to denote all positive relationships between two individual persons.[28]

Friendship is integral to the being of humans in terms of Christian theology. Augustine was the first Christian theologian who spelled out a theological notion of friendship: 'a true friendship ... is not possible unless [God bonds] together those who cleave to one another by the love which "is poured into our hearts by the Holy Spirit who is given to us" (Rom. 5:5).'[29] Indeed, Augustine's distinctively Christian concept of friendship will throw up questions about the value of the friendship between the non-Christian. However, he rightly flags up how friendship is bound up with the being of humans. In his position, the being of humans transformed by God's salvation can obtain the genuine friendship.

Augustine's awareness of the ontological implication of friendship had yet to be fully developed. On this front, Thomas Aquinas offers a more compelling theological exposition of friendship as integral to the human being, which begins with God–human relationship.

> [S]ince there is a communication between man and God, inasmuch as He communicates His happiness to us, some kind of friendship must needs be based on this same communication, of which it is written (1 Cor 1:9): *God is faithful: by Whom you are called unto the fellowship of His Son*. The love which is based on this communication, is charity: wherefore it is evident that charity is the friendship of man for God.[30]

The main point Aquinas makes in this passage is the friendship between God and humans established by virtue of God's love. In addition, the *communication* (*communicatio*) between humans and God is the means by which God–human friendship is actualised. In Aquinas's work, the communication between God and humans is conceptually associated with God's creation, whereby God communicates his own likeness or image, perfection, and goodness.[31] For this reason, the God–human friendship is the ultimate exemplar of all friendship between humans, which even extends to their enemies.[32]

While Augustine is preoccupied with Christian friendship generated through God's salvation, Aquinas adopts the doctrine of creation to open up a possibility to articulate a theology of friendship in general. Following Aquinas's line of thought, the holistic interpretation of the *imago Dei* and the thinking of archetype–ectype mean that the human being created *as* the *imago Dei* is embedded with the communicated love of friendship. For this reason, the love of friendship is part of the human being as a whole and is the *ectype* of the *archetypal* divine love of friendship—that is, the love between the three persons of the Trinity. From this we can infer two implications for human friendship. First, inasmuch as friendship is part of *psychosomatic* humanity, both the human mind and the body should play a role in making, performing, and maintaining friendship. Second, the archetypal friendship between the Father, the Son, and the Spirit is the ultimate model of all friendship. One crucial hallmark of the archetypal friendship is the distinction between the three divine persons. Each divine person is the other to the other two, and otherness is intrinsic to the unity of the three divine persons. By the same token, otherness is inherent to friendships between human persons. This point will be highlighted and reiterated in the below discussion on the distinction between human and artificial companionship.

These two implications are applied to and even become bolder in the romantic relationship between humans. If human friendship is the ectype of the love of friendship between the three divine persons, the romantic love between humans more vividly mirrors the love intrinsic to the inner life of the Trinity. In this respect, Karl Barth's work presents us with a cogent analysis of the *imago Dei* in relation to the romantic relationship. In Section 1.1 of Chapter 1, I pointed out that Barth's relational reading of the *imago Dei* is, for Herzfeld, conducive to interaction between AI and humans. Yet, Barth's interpretation of the *imago Dei* is more than relational.

> "He created them male and female." This is the interpretation immediately given to the sentence "God created man." As in this sense man is the first and only one to be created in genuine counterpart (*Gegenüber*) to God and as a genuine counterpart to his fellows—[to be created in genuine, actual otherness (*Andersheit*)]; it is he first and alone who is created "in the image" and "after the likeness" of God....[The image of God] is not a quality of man.... It does not consist in anything that man is or does.... He willed the existence of a being (*Wesen*) which in all its non-deity and therefore its differentiation can be a real partner; which is capable of action and responsibility in relation to Him; to which His own divine form of life is not alien; which in a creaturely repetition, as a copy and imitation, can be a bearer of this form of life.[33]

Recent Barth studies demonstrate that the ontological term '*Wesen*' connotes 'being' and 'essence' in the sense of the nature of substance, which determines a thing as it really is.[34] Viewed in this light, the *imago Dei* as relationship is, for Barth, a *substance* of humanity, despite his claim that the *imago Dei* 'does not consist in anything that man is.' As such, Barth's interpretation of the *imago Dei* is both relational and substantive, albeit with a different meaning of 'substantive' in contrast to the conventional substantive interpretation.

With this relational-substantive notion of the *imago Dei* in mind, Barth stresses that both the otherness and the divine form of life (i.e., the inner relational life of the Trinity) are imitated among humans and particularly reflected in the romantic relationship between the male and the female.[35] In the analysis of the otherness and the romantic relationship, Barth underscores the physical alterity as manifested in the sexes and the differentiation of the male and the female.[36] Like friendship, therefore, the romantic relationship involves psychosomatic humanity. To take Barth's point of view a further step, it can be argued that a genuine romantic relationship is intertwined with both mental and bodily activities.

Considering companionship in light of the two types of interpersonal relationships, it is clear that humans accompany with another in their own otherness which is intrinsic to the being of humans. Both friendship and romantic relationships are grounded in the *imago Dei* as inherent to holistic humanity.

5.2.2 Companionship in Solidarity

Companionship consists not only of relationships between two individual persons but also of one's relationship with the members of a community where they belong to. By drawing on Bavinck's theology of the *imago Dei*, I have spelled out in Chapter 1 subsection 1.3.2 that the *imago Dei* cannot be fully understood unless we reckon with it in a collective sense.

> [J]ust as the cosmos is a unity and receives its head and master in humankind; and just as the traces of God (*vestigia Dei*) scattered throughout the entire world are bundled and raised up into the image of God of humankind; so also that *humanity in turn is to be conceived as an organism that, precisely as such, is finally the only fully developed image of God*. Not as a heap of souls on a tract of land, not as a loose aggregate of individuals, but as having been created out of one blood; as one household and one family, humanity is the image and likeness of God.[37]

Given that humans share the *imago Dei*, individual persons are united with one another and belong to the same organism of humanity. In this organic community, humans do not isolate themselves but rather accompany one another, which means that they are companions for one another. All human persons of this community act towards the flourishing and well-being of the organism of humanity. To this end, there is a bond that welds together all human beings for the sake of the interests of community members. It is this bond that can be designated as solidarity. Hence, the *imago Dei* is the ontological underpinning of human companionship, and communal solidarity is the exhibition of companionship within community.

The intertwinement between the collective implication of the *imago Dei* and communal companionship entails that solidarity should be construed in tandem with psychosomatic humanity. Like friendship and romantic relationship, solidarity is not only something that we feel through our consciousness. Furthermore, human biological bodies feature in solidarity, which comes to the fore in recent studies on evolution. I have touched upon niche construction theory in Section 1.2 of Chapter 1, pointing out the collective meaning of the *imago Dei* in light of human evolution. Added to this is Agustín Fuentes's studies on niche construction that draw our attention to psychosomatic humanity as the foundation of human solidarity. He argues that the construction of the human niche is inextricably linked with the changes of the bodies and the minds.[38] Whilst investigating human niche construction and culture, he underscores the psychosomatic feature of humanity:

> The specific patterns of how we use our bodies, minds, and skills are grown and incorporated into the human organism through practice and training in a given environment. They are thus simultaneously biological and cultural. Skills are contingent on the capacities and constraints of our bodies and our relationships with one another and the social, historical, and material environments we inhabit.[39]

As per Fuentes's finding, both human bodies and minds underpin human relationships, testifying to the evolution of psychosomatic human beings happened in human solidarity while constructing their niche. For example, human behaviours and bodies evolved in the course of cooperation on parenting and caregiving, which facilitated the advent and development of *compassion*, 'a central feature of the human niche.'[40] From this vantage point, it can be argued that companionship is interwoven into humanity through evolution of human beings in solidarity.

Inquiry into communal companionship and solidarity in light of the collective implication of the *imago Dei* should not leave aside the fact that a single human person as the *imago Dei* differs from other human fellows as the *imago Dei*. The communal unity of the *imago Dei* does not overturn the diversity of the individual *imago Dei*. This unity-in-diversity of the *imago Dei* signifies the important place of *otherness* in the development of communal companionship, demonstrating that the otherness of each human person is the ectype of the otherness intrinsic to the divine archetype. On this front, Dietrich Bonhoeffer's elaboration on otherness-in-community can help flesh out companionship in solidarity with reference to the particularity of individual persons. He argues that the I-Thou relationship between humans is grounded in the relationship between humans and God, that is, the relationship between the human I and the divine You. Through God's creation, the human You comes into existence and can be called the image of the divine You, which means that the human You is in no way absolute. For this reason, Bonhoeffer insists that the divine You as the absolute You 'becomes visible in the concrete You of social life.'[41] From this it follows that the human You as the other to the human I varies across social relations, showing the uniqueness of individual human persons. As such,

Bonhoeffer argues that 'in some way the individual belongs essentially and absolutely with the other, according to God's will, even though, or precisely because, the one is completely separate from the other.'[42] The paradox of 'belong with' and 'separate from' sheds light on companionship in communal solidarity: companionship characterised by solidarity presupposes otherness inherent to the community.

Companionship in solidarity unveils both how human companions belong within the human community and how the psychosomatic human being serves for the embodiment of the collective implication of the *imago Dei*.

5.2.3 Companionship Established by Empathy

Having analysed human companionship from the perspectives of both interpersonal relationships and solidarity, two crucial features of companionship come to the fore. First, friendship, romantic relationship, and communal solidarity presuppose psychosomatic humanity, which means that both human physical and mental particularities are of equal importance for companionship. In this regard, the whole human person as the *imago Dei* becomes salient through companionship. Yet, a question arises: How does companionship take shape in the unity of the body and the mind? Second, companionship is always concomitant with otherness. To be a companion is, first of all, to be an other to someone else, be it a friend, a partner, or group members. When companionship is formed, otherness is often overshadowed. This feature is in accord with the collective implication of the *imago Dei*. That is, the meaning of the image of God cannot be fully comprehended unless through the lens of human community. The second feature of companionship leaves us with another question: How does *otherness* gradually step into background and yet remain to exist throughout the formation of companionship?

These questions are tied up with the way in which the psychosomatic human being acts as an other to human fellows. I will take a further step to inquire into *how* psychosomatic humanity works to create, sustain, and facilitate companionship while maintaining otherness. I will concentrate on the role that empathy plays in human companionship. Generally speaking, empathy means to put ourselves into someone else's shoes, showing feelings and experiences shared between us and others. Therefore, otherness is a precondition for empathy. Both interpersonal and communal relationships are partly yet significantly predicated upon empathy. Companionship in friendship, romantic relationship, and communal solidarity alike presupposes shared *feelings and experiences*, which are part of human consciousness. In this light, empathy implies the resonance between human consciousnesses.

Both consciousness and companionship proceed from holistic humanity, which means that empathy is closely interwoven with the human being as a whole. Shannon Vallor presents us with a vivid portrayal of an empathic human person.

> Empathy may include sharing feelings of pain, fear, shame, excitement, joy, or other passions. Though we can speak of purely cognitive empathy (i.e., the correct belief that another is experiencing a certain emotion), empathy is often associated with bodily passivity, an experience of being physically *affected* by another's emotion. It is an intimate, personalized form of relation; I can have sympathy for the misfortunes of a group of anonymous individuals who are more or less interchangeable in my mind, but I typically can only empathize with a specific person or set of persons. Even a stranger with whom I empathize will appear in my experience not as an abstraction, but as a concrete, nonfungible individual with specific emotional experiences.[43]

A crucial point made in this passage is that empathy with shared feelings and emotional experiences is the consequence of the cooperation of the mind and the body but, more importantly, must come into existence through human bodies. But how does the human body act as a catalyst for the generation of empathy?

Recent neuroscientific studies offer important insights into the function of the body in empathy. Since empathy is concomitant with feelings, Antonio Damasio's neuroscientific investigation into feeling brings forth a helpful account of how human bodies play a part in the generation of empathy. According to him, it is through nervous systems that feelings are inextricably linked to human bodies.

> Feelings owe their existence to the fact that the nervous system has direct contact with our insides and vice versa. The nervous system literally "touches" the organism's interior, everywhere in that interior, and it is "touched" in return. The nakedness of the interior relative to the nervous system and the direct access the nervous system enjoys relative to that interior are part of the uniqueness of interoception, the technical term reserved for the perception of our visceral interior.[44]

Through the human nervous system, feelings bring together the mind and the body in such a way that the state of the body is represented in the mind.[45] As such, Damasio argues: 'Feeling provides us with knowledge of life in the body and, without missing a beat, makes that knowledge conscious.'[46]

Considering empathy in conjunction with Damasio's view of feeling, empathy involving shared feelings carries two implications. First, empathy means that the state of another's body is shared and represented in our minds through feelings as if our bodies are in the same state. The similarity of the two conscious representations of the state of the body makes companionship possible and facilitates the development of interpersonal and communal relationships during human niche construction.[47] Furthermore, the similar representations of the states of bodies imply the otherness between the empathiser and the recipient of empathy: 'the empathizer has interoceptive awareness of his or her own bodily states and is able to distinguish between self and other.'[48] As such, empathy is a *condicio sine qua non* for companionship in otherness.

Second, and added to the first implication, neurons and neural networks underpin feelings. It is the second implication through which we can discover how the brain, as the specific part of the body, is a factor in the generation of empathy. Neuroscientific studies have demonstrated that the brain matters to empathy. Cristina Gonzalez-Liencres and her colleagues show that empathy is indexed to neurochemicals, that is, organic molecules for neural activities. For example, empathy and pro-social behaviour are associated with the increase of the neuropeptide oxytocin.[49]

In addition to neurochemicals, recent studies on the neural basis of empathy put the spotlight on the neural mechanisms of empathy with the aid of functional neuroimaging—that is, using technologies to discover the association between mental functions and the specific areas of the brain. For example, neuroimaging research on pain and empathy proves that the same brain regions of the empathiser and the recipient of empathy are activated when the former empathises with the latter. Hence, the empathiser feels an empathic or vicarious pain.[50] It has been well demonstrated that the mirror neuron mechanism underlies empathy in light of a perception–action model. In subsection 3.3.2.2 of Chapter 3, I have spelt out mirror neuros, which refer to a subset of neurons for understanding and imitating the actions of others. The perception–action model is raised by Stephanie Preston and Frans de Waal, which means that 'attended perception of the object's state automatically activates the subject's representations of the state, situation, and object, and that activation of these representations automatically primes or generates the associated autonomic and somatic responses, unless inhibited.'[51] With the mirror neuron mechanism, the empathiser perceives the other's state and consequently feels a similar emotion.[52]

Neuroscientific studies on empathy demonstrate how the mind and the body together serve as the infrastructure of empathy, which, in turn, leads to companionship. It is the psychosomatic human being who is conscious of becoming a companion when empathising with others. That said, it is worth highlighting that neuroscience does not materialise or biologise empathy. Rather, social environment and culture have a bearing on empathy, corroborating the fact that empathy is dynamic and can be cultivated as a virtue over the time.[53]

Taking the above analysis together, it can be argued that empathy lies at the centre of all companionship. Empathy illustrates how psychosomatic humanity is displayed through friendship, romantic relationship, and communal solidarity whilst maintaining otherness in companionship. From this perspective, it is evident that the human individual as the *imago Dei* becomes a companion of another *imago Dei* and also serves for the collective implication of the *imago Dei*.

5.3 An Ectypal Account of Companionable AI

The preceding two sections have respectively examined the concept of companion from the perspectives of AI and theology (in conjunction with neuroscience), providing a lens through which to explore how theology can help to address issues related to the application of AI to human companionship. From the vantage point of the holistic interpretation of the *imago Dei*, it is unequivocable that none of the substantive, the functional, and the relational approaches to artificial companion can do justice to holistic humanity. As such, it is expected that the theology of the *imago Dei*, along with the thinking of archetype–ectype, can serve to reshape a hermeneutical approach to the deployment of AI in companionship.

5.3.1 Going Beyond the Three Hermeneutical Approaches to Artificial Companion

It should be acknowledged that the substantive, the function, and the relational understandings of artificial companions provide us with helpful insights into human companionship and remind us of crucial factors in establishing companionship. Be that as it may, a more integrative hermeneutical tool needs to be developed to address the issues that remain unsolved by the three approaches.

Seeking a new hermeneutical approach should begin with an investigation into the problems inherent to these three understandings of artificial companion. Based on the above theological analysis of companionship, I proceed to examine two issues: (1) the ELIZA effect on companionship, and (2) the vanishing of otherness.

The ELIZA effect is named after the chatbot ELIZA created by Joseph Weizenbaum with early natural language processing technology in 1966 at the Massachusetts Institute of Technology.[54] This effect refers to a phenomenon that human users are inclined to attribute human capacities, such as emotions, intelligence, and reciprocity, to interactive computational artefacts like ELIZA, that is, artificial companions. It reflects 'the susceptibility of people to read far more understanding than is warranted into strings of symbols—especially words—strung together by computers.'[55] The substantive and relational understandings of artificial companion are resonant with the ELIZA effect. Luciano Floridi's substantive interpretation of artificial companion turns out to be the humanisation of AI-powered artefacts, and the relational portrayal of artificial companion like David Levy's account of artificial lover attributes human-level relationality to computational artefacts.

Theologically speaking, the ELIZA effect is a compelling evidence to showcase that the human being as the *imago Dei* is the created co-creator who brings the *imago hominis* into existence (see Chapter 1, Section 1.4). As such, the concept of *imago hominis* theologically justifies the possibility of using AI to advance human companionship. (It will be fleshed out in subsection 5.3.2.) Be that as it may, the ELIZA effect lays bare the impossibility of human-level artificial companion precisely because the human capacities for companionship are *attributed* to artificial companion rather than *fully reproduced* in AI systems. The radical distinction between the *imago hominis* and the *imago Dei* continues to serve as a foundation for the uniqueness of human companion vis-à-vis AI.

The second issue related to the three hermeneutical approaches is the vanishing of otherness in the companionship between humans and AI. Apart from Floridi's human-level artificial companion, the substantive interpretation of artificial companion may identify an artificial companion as a slave and a tool to accompany humans for our well-being. This slavery position also ties in with the functional interpretation of artificial companion, which may pay little or no attention to moral concerns about AI *per se*.

Otherness also vanishes in the relational interpretation of artificial companion. For an AI-powered interactive artefact is considered an independent entity that is operating to provide humans with company. In fact, there is an illusion: the relationship between the human user and the artificial companion as the other is, in substance, the relationship between self and an *artificial self*. Otherness vanishes in self-companionship because artificial companions are designed according to the personal needs of human users. Again, David Levy's artificial lover is illustrative here: 'two people are more likely to fall in love with each other if they are similar in some respects; so *a computer endowed with knowledge about a user's hobbies, and programmed with similar personality traits to the user*, will again have a head start in its attempts to engender love from its human.'[56] It is apparent that the tendency to make a replicate of self in the artificial lover implicitly rules out the possibility of otherness. Sherry Turkle offers a perceptive observation on this point: 'The first thing missing if you take a robot as a companion is alterity, the ability to see the world through the eyes of another. Without alterity, there can be no empathy.'[57] Because of artificial companions, human users are caught up in self-interaction and self-desiring.

The inquiry into these two issues forces us to reconsider the concept of artificial companion. Is the term 'artificial companion' *in se* conceptually justified? In fact, blurred boundaries between artificial and human companions are largely resulted from the conceptual ambiguity of companion, which is at variance with the real-life experience of companionship between humans. Julia Knifka's assertion is spot-on.

> A social interaction consists of more than a mutual observation. The interaction partners have to be aware of the other and his (biographically determined) situation and they have to have the ability to react to someone. A robot might have the ability to respond to me and recognize behavioral patterns, but a true social interaction will not be possible, unless I can see myself in the other and I know the other can take my perspective and will be able to understand me.[58]

Given this self-consciousness of our mutual otherness in conjunction with the consciousness of the other as a whole human person, I suggest that 'companionable AI' is an ideal terminological substitute for 'artificial companion' for two reasons. First, 'companionable AI' orients us to shift the focus from 'companion' to AI itself, implying the ontological distinction between artificial and human companions. At the same time, such a shift of the focus helps us to tease out the specific role of AI in fostering companionship. Second, 'companionable AI' reminds us of how and for what purpose AI is created to be companionable. Now we are turning to the articulation of companionable AI with the aid of both the theo-ontology of AI and AI ethics.

5.3.2 *Companionable AI: An Interdisciplinary Approach*

The theology of *imago Dei* and the thinking of archetype–ectype together lend theological credence to the concept of companionable AI. Since companionship is a constituent of the *imago Dei*, AI as the *ectype* of the image of God should, to a certain extent, be companionable insofar as the human being as the ectype of God acts as the created co-creator to bring companionable AI into existence. As such, companionable AI can be considered the ectype of humanity, being distinguished from human companions and intended to collaborate with humans to foster companionship between human users. From this, two implications arise. First, companionable AI as the *ectype* of human companions carries the *eccentricity* of the *imago Dei*, which connotes that the meaning of companionable AI is grounded outside itself in specific ways in which human companions relate to one another. Second, companionable AI is inbuilt with the potential to produce moral impacts on human companionship. The concept of companionable AI does not fit into the categories of the substantive, the functional, and the relational approaches to artificial companion; in fact, it goes beyond the three hermeneutical approaches and offers a more realistic account of the real-life applications of AI to human companionship, which paves the way for the inclusion of companionable AI into human communities.

This theologically reconfigured concept of companionable AI is resonant with some major elucidations on human–AI interaction. Sherry Turkle alerts us to an illusion of companionship provided by computational artefacts.

> But if our experience with relational artifacts is based on a fundamentally deceitful exchange (they perform in a way that persuades us to settle for the "acting out" of caring), can they be good for us? Or, as I have asked, might they be good for us only in the "feel good" sense? The answers to such questions do not depend on what computers can do today or are likely to be able to do tomorrow. They depend on what *we* will be like, the kind of people *we* are becoming as we launch ourselves and those we love into increasingly intimate relationships with machines.[59]

Following Turkle's logic, due attention should be given to the asymmetry between the human being as the archetype and companionable AI as the ectype, emphasising the dominant place of ourselves in AI-powered companionship. All the more important is that Turkle sheds light on the way in which companionable AI belongs into human communities, aligning with Paul Dumouchel and Luisa Damiano's envisioning of future interaction between humans and robots. Dumouchel and Damiano are convinced that 'a close relationship with technological objects is an occasion for moral and psychological growth' insofar as '[l]iving with robots holds out the prospect of a better future—not only, and not primarily, a more prosperous future in the economic sense, but a better future morally, a more humane future.'[60] But how can companionable AI contribute toward a better future? How can companionable AI play its due role in the development of human well-being while not replacing human companions altogether?

Whilst addressing these two questions, we should develop a nuanced understanding of the various applications of AI in fostering companionship, as the term 'companion' is construed in a very broad sense. On the whole, the self-centred application should be differentiated from the rest use of AI for companionship. By 'self-centred,' I refer to the AI-powered companionable artefact that is designed as the substitute for human companions to meet the needs of human users. In the final analysis, self-centred companionable artefacts are nothing other than attempts made to close the ontological gap between the *imago Dei* and the *imago hominis*. An example of this type is the artificial lover. Charles Ess puts forward a strong critique of the project of artificial lover.

> However skills and pleasing my robot lover might be—as an entity incapable of awareness and desire, it cannot be aware of desiring me, much less of desiring my desire. In turn, it can never give me the vital experience of knowing that it is an Other who desires my desire. Absent these, it cannot provide me with the fullest affirmation of my sense of self as a unique person, as this selfhood includes my desire that my desire be desired.[61]

Accordingly, self-centred AI-powered companionable artefacts will eventually undermine human well-being. In contradistinction to these artefacts, companionable AI is created to collaborate with humans in fostering human companionship.

Studies on building empathy into AI undergird companionable AI as the ectype and collaborator of human companions.[62] Note, however, that empathic AI is not made to create human-level artificial companions or self-centred AI-powered artefacts in order to provide whatever human users desire for. Rather, companionable AI embedded with empathy is designed to co-work with humans in eliciting empathic responses from humans. Since there are many things and many hands involving in the creation of AI systems (see Chapter 4, subsection 4.3.1), various factors should be taken into consideration while developing such empathic AI. On this front, Ramya Srinivasan and Beatriz González's exposition is compelling:

> The incorporation of empathy in AI systems calls for creation and enforcement of new capabilities and requirements such as to consider subjective viewpoints of different stakeholders, to base decisions not only on proxy data, and to appeal across stakeholders. These requirements necessitate an understanding of all stakeholders' intentions and contextual information related to their affective states, intentions, and socialization patterns. Furthermore, empathetic AI systems should care for users' well-being, and demonstrate responsibility in decision making.[63]

Accordingly, empathic AI should integrate the empathy of AI designers, programmers and stakeholders with the needs and benefits of human users. In this vein, companionable AI becomes a medium that intertwines multiple strands of empathy with those who demand companionship. In other words, companionable AI is the empathic *imago hominis* that mediates and extends the companionship between one *imago Dei* and another. As such, companionable AI augments the human being's inherent capability to accompany others. It is particularly in this sense that companionable AI retains otherness in human companionship and shows its purpose to contribute toward the flourishing of humans. Due to the limited space, I cannot go into a comprehensive analysis of the applications of companionable AI to our lives. In what follows, AI-powered caregiving practices and carebots will be utilised to illustrate companionable AI as the *imago* of human caregivers, mindful of Christian pastoral care.

It has been commonly accepted that carebots exert an impact on human caregivers. Shannon Vallor asserts that 'carebots might deprive potential caregivers of important moral goods central to caring practices' and, at the same time, 'might help caregivers sustain or even enrich those practices, and their attendant goods.'[64] She points out, for example, that carebots can help us extend our empathic practices towards others but, at the same time, may make us inert in the development of empathy. For in performing empathic practices, humans are conscious of and respond appropriately to the suffering and pains of another, which facilitates the cultivation of their empathy with those who suffer.[65]

Empathic caregiving practices reflect the reciprocity that is intrinsic to companionship between humans. Reciprocity means giving and receiving among humans when they are doing or having done for one another, which indicates that human–human interactions like companionship and caregiving are bidirectional rather than unidirectional. As Vallor asserts, 'We learn in caregiving practices to see how reciprocity holds human relations together and allows other kinds of goods to flow across them.'[66] Aimee Van

Wynsberghe, a leading expert in the field of AI ethics, unpacks the importance of reciprocity for caring relationship.[67] (The same holds true for other types of human companionship.) She locates reciprocity at the centre of the moral value of all technology:

> the technology cannot be evaluated according to its impact on either caregiver or care receiver, *instead it must be evaluated according to its impact on the ability to establish a reciprocal relationship throughout the practice*.... In the absence of a moral priority given to the caring relationship, where that is the first and essential item to preserve, then the likelihood is that humanitarian robots will be built with narrower values in mind: keeping people alive and uninjured, as cheaply as possible.[68]

Accordingly, carebots as companionable AI should be designed to consolidate reciprocity between humans lest the moral value and impact of technology is downplayed or ignored.

Reciprocity is more complicated than bilateral relationship when looking at specific caring relationship in terms of companionship. Companionable carebots for reciprocal interpersonal relationship are always associated with particular human persons, including, among others, their specific social setting, cultural background, and personalities. In like manner, companionable carebots for reciprocal communal relationship are consistently relevant to a particular community and should engage with its tradition, ethos, and value. These characteristic features of carebots for reciprocal companionship echo the concept of contextually confined moral agency (see Chapter 4, subsection 4.2.2).

Having this reciprocal nature of companionship in mind, Van Wynsberghe goes on to develop a methodology for the design of carebots, which is named 'Care-Centred Value-Sensitive Design' (CCVSD).[69] She suggests that CCVSD begins with discussion over care practices. Following this, she argues:

> The ethical analysis of the relationship between robot capabilities and its impact on the manifestation of moral elements, roles and responsibilities is repeatedly debated throughout the design process right up to the implementation of the robot. In this way ethics (i.e. ethical value analysis) is included throughout the entire process.[70]

Van Wynsberghe is convinced that CCVSD will help to create carebots *with* care. The phrase 'with care' carries two connotations. Firstly, the design of carebots should be embedded with the acknowledgement of the bearing carebots will have on human caregiving practices. Secondly, the technical content of carebots is intentionally inbuilt with care values (e.g., compassion, respect, accountability) through designing process.[71]

Taking these together, Van Wynsberghe maintains that the design of carebots must attend to 'the specific context of use, the unique needs of users, the tasks for which the robot will be used, as well as the technical capabilities of the robot.'[72] Based on CCVSD, she proposes a care-centred framework for the evaluation of whether or not carebots achieve their goals. This framework has five components: context, practice, actors involved, type of robot, and manifestation of care values.[73] Caregiving practices varies across different contexts. Without doubt, a nursing home will have different caring practices than that in a hospital. The differing contexts give rise to different caregiving practices, such as bathing, delivery of medications, and chatting. The contextual and behavioural factors in caregiving practices are bound up with many human agents, particularly caregivers, care-receivers and the end users of carebots. The carebot is the junction both where human actors involved in care practices meet one another and where human companionship is performed. Hence, the different roles of human actors in caregiving practices should be reckoned with throughout the process of design. Given these divergent purposes and functions of care practices, carebots should be designed differently so that they can be used to accomplish specific tasks. These four components of the care-centred framework together stimulate the manifestation of moral elements and care values inherent to defined care practices. Van Wynsberghe designates four care values as fundamental to all care practices:

Attentiveness: capability of recognizing the dynamic needs of a patient;
Responsibility: closely aligned with trust; it requires an understanding of what one is doing and why; capability of identifying response to needs and delegation to meet them;
Competence: capability of executing a means/action to fulfil the identified needs in a skilled manner;
Responsiveness: capability to engage with the care-receiver regarding the meeting of their needs.[74]

These four components should be foregrounded at the beginning of the design of carebots in order that reciprocity can be established between caregivers and care-receivers. Van Wynsberghe's CCVSD is in line with and instantiates the concept of companionable AI in that the care-centred framework concentrates upon AI itself and foregrounds the purposes, design, and functions of carebots. What is even more crucial is that the CCVSD and the derived care-centred framework present us with an example to illustrate how companionable AI as the *imago hominis* can serve for the companionship between one *imago Dei* and another.

Van Wynsberghe's CCVSD is directed at carebots in healthcare, but it is a heuristic for considering the deployment of carebots in Christian pastoral care. By Christian pastoral care, I refer to the care that Christian clergy, communities, and laity offer from a theological perspective to those who are undergoing either mental or physical suffering and troubles or both. David Lyall teases out some important elements of Christian pastoral care.

> Pastoral care involves the establishment of a relationship or relationships whose purpose may encompass support in a time of trouble and personal and/or spiritual growth through deeper understanding of oneself, others and/or God. Pastoral care will have at its heart the affirmation of meaning and worth of persons and will endeavour to strengthen their ability to respond creatively to whatever life brings.[75]

In Lyall's position, it is clear, both the being of humans and their relationality lie at the centre of pastoral caregiving practices, and human relationship will be enhanced through pastoral care of the being of humans. Given our holistic interpretation and the religious nature of the *imago Dei*, Christian pastoral care can be considered an excellent illustration of companionship *between* individual *imago Dei* and *within* the community of the *imago Dei*.

A note has been made at the beginning of this study that some theologians have examined questions related to AI-powered pastoral care. William Young reminds us that religious communities should be ready to deal with the questions on the reception of 'automating relationships in ministry' because of the rapid progress in technology.[76] To be sure, AI-driven systems can liberate church ministers from some routine work of pastoral care. For example, some Christian believers may expect ministers to send out Bible verses every day so that they can be strengthened to go through and endure occasional troubles and difficulties. We can imagine that an AI-driven automated system is capable of sending daily Bible verses that responds to one's troubles or to topical events that are likely to trouble religious believers (e.g., the COVID-19 pandemic). However, this liberation from some routine work does not mean AI-powered pastoral carebots replace human ministers. The routine work of pastoral care is constantly followed by further meticulous and detailed caregiving practices (e.g., intercessory prayer and empathic response), which, according to the idea of companionable AI, cannot be accomplished by pastoral carebots. As such, pastoral carebots are not made to undermine human pastoral skills but, to put it in Eric Stoddart's words, '[address] barriers to people's participation.'[77] In this way, ministers can focus more attention on others' critical needs of pastoral care, say, pastoral care at the end of life. Moreover, an AI-powered pastoral carebot is a form of companionable AI that is designed to cultivate reciprocal relationships between humans through caregiving practices.

Pastoral carebots created in the manner of companionable AI follow Van Wynsberghe's CCVSD and fit in well with the care-centred framework. That is to say, pastoral carebots are designed to perform and *augment*

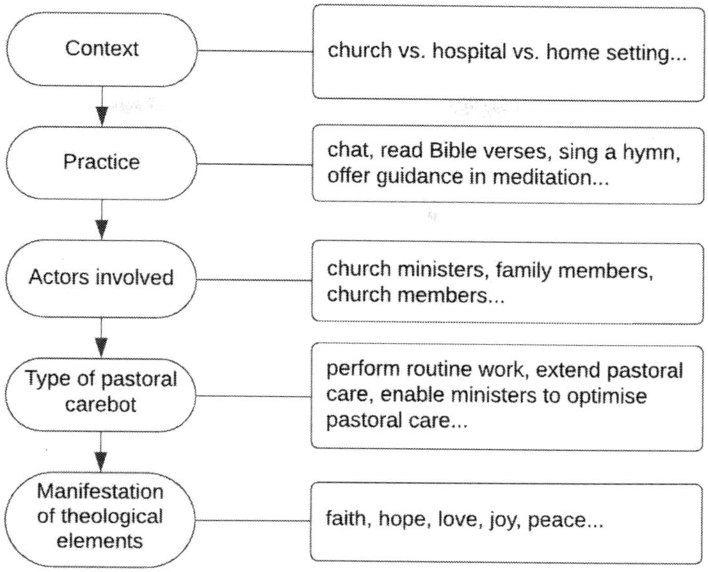

Figure 5.1 Pastoral-care-centred framework

caregiving practices after the model of human pastoral caregivers. Hence, Van Wynsberghe's care-centred framework can be reconfigured as shown in Figure 5.1.

The first four components are largely similar to those of Van Wynsberghe's care-centred framework. Attention needs to be drawn to the fifth component of the framework. I replace her 'moral elements' with 'theological elements.' However, her four moral elements and care values (i.e., attentiveness, responsibility, competence, responsiveness) remain effective in pastoral-care-centred framework but are expected to become manifest through theological elements. For example, the Christian notion of hope can be a value that is used to evaluate pastoral carebots. AI designers and the policymakers of Christian communities can investigate whether care-receivers in a specific pastoral context hold the hope for God's faithfulness and deliverance more firmly with the addition of carebots to pastoral care. To accomplish this task, pastoral carebots should recognise the dynamic religious needs of care-receivers (attentiveness), understand their disappointment or despair and the causes thereof (responsibility), perform actions to strengthen their hope (competence), and engage with their responses regarding hope (responsiveness). In this vein, the theological elements of Christian pastoral carebots actualise Van Wynsberghe's care values.

This proposed framework can be used to guide the design process of pastoral carebots in a specific way and for particular caregiving practices.

The five components should be discussed and weighed up time and again throughout the design process of Christian pastoral carebots such that pastoral carebots can co-work with (rather than substitute for) humans to cultivate companionship between one another. By doing so, companionable AI is made capable to participate into human ethical performance towards others.

5.4 Conclusion

How can AI, as the *imago hominis*, participate in the ethical performance of the *imago Dei*? To answer this question, I have probed the concept of artificial companion in this chapter, exploring the substantive, the functional, and the relational interpretations. Following this, it has been demonstrated that the theological construal of holistic humanity as the *imago Dei* offers us with a more constructive and compelling approach to the application of AI to human companionship, leading to a promising notion of companionable AI as the *imago hominis*.

Furthermore, AI as the ectype of humanity provides a solid theo-ontological foundation for companionable AI. That is, AI is not and cannot be designed to replace human companions altogether in that the ontological distinction between AI and humans undergirds the uniqueness of humans in companionship compared to companionable AI. With the aid of the thinking of archetype–ectype, companionable AI reflects the eccentricity of the *imago Dei* and, consequently, is closely connected to human companionship. Stanford University's AI100 Study Panel Report (2021) reviews recent applications of AI technology to caregiving practices and draw a perceptive corollary.

> The use of AI technologies in caregiving should aim to supplement or augment existing caring relationships, not replace them, and should be integrated in ways that respect and sustain those relationships. Good care demands respect and dignity, things that we simply do not know how to code into procedural algorithms. Innovation and convenience through automation should not come at the expense of authentic care.[78]

In similar manner, human companionship demands many more things that are unable to be mathematised and coded into AI. Companionable AI helps us to demystify artificial companion and seek for collaboration between humans and companionable AI to cultivate and augment companionship between humans.

Notes

1 Yorick Wilks, "Foreword," in *Close Engagements with Artificial Companions: Key Social, Psychological, Ethical and Design Issues*, ed. Yorick Wilks (Amsterdam: John Benjamins, 2010), xi.

2 Jutta Weber, "Opacity versus Computational Reflection: Modelling Human-Robot Interaction in Personal Service Robotics," *Science, Technology & Innovation Studies* 10, no. 1 (2014): 192–194.
3 Luciano Floridi, *The Fourth Revolution: How the Infosphere is Reshaping Human Reality* (Oxford: Oxford University Press, 2014), 87–90.
4 Ibid., 93.
5 Turing, "Computing Machinery and Intelligence," 433–460.
6 Floridi, *The Fourth Revolution*, 93–94; emphasis added.
7 Ibid., 96.
8 Luciano Floridi, "Artificial Companions and Their Philosophical Challenges," in *Close Engagements with Artificial Companions: Key Social, Psychological, Ethical and Design Issues*, ed. Yorick Wilks (Amsterdam: John Benjamins, 2010), 27; Luciano Floridi, "Artificial Intelligence's New Frontier: Artificial Companions and the Fourth Revolution," *Metaphilosophy* 39, no. 4/5 (2008): 654.
9 Joanna J. Bryson, "Robots Should Be Slaves," in *Close Engagements with Artificial Companions: Key Social, Psychological, Ethical and Design Issues*, ed. Yorick Wilks (Amsterdam: John Benjamins, 2010), 65.
10 Ibid., 73.
11 Aaron Sloman, "Requirements for Artificial Companions: It's Harder than You Think," in *Close Engagements with Artificial Companions: Key Social, Psychological, Ethical and Design Issues*, ed. Yorick Wilks (Amsterdam: John Benjamins, 2010), 182.
12 Cora Lydon, "Edinburgh Hospital Trials AI-powered Pain Assessment Tech," DigitalHealth, 21 December 2023, accessed 30 December, 2023, https://www.digitalhealth.net/2023/12/edinburgh-hospital-trials-ai-powered-pain-assessment-tech/.
13 For more examples of artificial companions for healthcare, see Chee-Peng Lim et al., eds., *Handbook of Artificial Intelligence in Healthcare, Vol. 1—Advances and Applications* (Cham: Springer, 2022).
14 On further, see Joelle Pineau et al., "Towards Robotic Assistants in Nursing Homes: Challenges and Results," *Robotics and Autonomous Systems* 42 (2003): 271–281.
15 Kari Gwen Coleman, "Android Arete: Toward a Virtue Ethic for Computational Agents," *Ethics and Information Technology* 3 (2001): 248.
16 Ibid., 257.
17 David J. Gunkel, *The Machine Question: Critical Perspectives on AI, Robots, and Ethics* (Cambridge, Massachusetts: MIT, 2012), 87. His intention is to recast the term 'personhood' in the age of AI (88–90).
18 Ali Sekmen and Prathima Challa, "Assessment of Adaptive Human-Robot Interactions," *Knowledge-Based Systems* 42 (2013): 49.
19 Amanda Sharkey and Noel Sharkey, "Granny and the Robots: Ethical Issues in Robot Care For the Elderly," *Ethics and Information Technology* 14, no. 1 (2012): 27–40; also see Noel Sharkey and Amanda Sharkey, "Living with Robots: Ethical Tradeoffs in Eldercare," in *Close Engagements with Artificial Companions: Key Social, Psychological, Ethical and Design Issues*, ed. Yorick Wilks (Amsterdam: John Benjamins, 2010), 245–255. It is worth nothing that Sharkey and Sharkey employ a narrower notion of companionship than that raised in this chapter. As per them, companionship. But their study fits in with my exploration of artificial companion in its broader sense. In addition to these six ethical concerns, there are other ethical issues related to the application of artificial companions. For example, some scholars discuss ethical issues related to brain and mental healthcare, see David D. Luxton and Eva Hudlicka, "Intelligent Virtual Agents in Behavioral and Mental Healthcare: Ethics and Application Considerations," in *Artificial Intelligence in Brain and Mental Health: Philosophical, Ethical &*

Policy Issues, ed. Fabrice Jotterand and Marcello Lenca (Cham: Springer, 2021), 53–55; Gary Marchant and Lucille Nalbach Tournas, "Matter Over Mind: Liability Considerations Surrounding Artificial Intelligence in Neuroscience," in *Artificial Intelligence in Brain and Mental Health: Philosophical, Ethical & Policy Issues*, ed. Fabrice Jotterand and Marcello Lenca (Cham: Springer, 2021), 233–246; Mónika Sziron, "A Common Ground for Human Rights, AI, and Brain and Mental Health," in *Artificial Intelligence in Brain and Mental Health: Philosophical, Ethical & Policy Issues*, ed. Fabrice Jotterand and Marcello Lenca (Cham: Springer, 2021), 247–258.

20 Sherry Turkle, *The Second Self: Computers and the Human Spirit*, 20th Anniversary ed. (Cambridge: MIT Press, 2005), 288–297.

21 David Levy, *Love and Sex with Robots: The Evolution of Human-Robot Relationships* (New York: HarperCollins, 2008), 22; emphasis added. Levy's work is influenced by Sherry Turkle's *The Second Self*. But interestingly, Turkle admits that she was 'underwhelmed by the message of … *Love and Sex with Robots*.' Sherry Turkle, *Alone Together: Why We Expect More from Technology and Less from Each Other* (New York: Basic Books, 2011), 5. For Turkle's critical response to Levy's work, see *Alone Together*, 5–7.

22 Levy, *Love and Sex with Robots*, 268–272; also see David Levy, *Robots Unlimited: Life in a Virtual Age* (Wellesley, Massachusetts: A K Peters, 2006), 315–346.

23 Levy, *Love and Sex with Robots*, 306–307.

24 Ibid., 144–150; also see David Levy, "Falling in Love with a Companion," in *Close Engagements with Artificial Companions: Key Social, Psychological, Ethical and Design Issues*, ed. Yorick Wilks (Amsterdam: John Benjamins, 2010), 89–93.

25 Levy, *Love and Sex with Robots*, 150.

26 Scott A. Midson, "From *Imago Dei* to Social Media: Computers, Companions and Communities," in *Love, Technology and Theology*, ed. Scott A. Midson (London: T&T Clark, 2020), 143–160.

27 Ibid., 159.

28 It is worth noting that the filial relationship is not part of interpersonal relationships. In terms of its origin, the filial relationship must involve three persons: the father, the mother, and the child. Hence, the filial relationship is a type of communal relationship.

29 Augustine, *Confessions*, trans. Henry Chadwick (Oxford: Oxford University Press, 1998), 4.4.7.

30 Thomas Aquinas, *Summa Theologiae*, trans. Fr. Laurence Shapcote, ed. The Aquinas Institute, Latin/English Edition of the Works of St. Thomas Aquinas, Vols. 13–20 (Green Bay, WI: Aquinas Institute, 2012), II-II, q.23, art. 1, resp.

31 Ibid., I, q.44, art 4.

32 Ibid., II–II, q.23, art.1, ad. 2–3.

33 Karl Barth, *Church Dogmatics, Volume III: The Doctrine of Creation, Part 1*, trans. J. W. Edwards, O. Bussey, and H. Knight, ed. Geoffrey William Bromiley and Thomas F. Torrance (London: T&T Clark, 2004), 184–185; revise translation; Karl Barth, *Die kirchliche Dogmatik, Band III,1* (Zürich: Theologischer Verlag Zürich, 1980), 206–207.

34 '*Wesen*' differs from '*Sein*' in that the latter refers to 'being' in the sense of the essence of a thing as it appears. Further on this, see Shao Kai Tseng, *Barth's Ontology of Sin and Grace: Variations on a Theme of Augustine* (London: Routledge, 2019), 48–75; Ximian Xu, "Karl Barth's Ontology of Holy Scripture Revisited," *Scottish Journal of Theology* 74, no. 1 (2021): 26–40.

35 Barth, *III/1*, 185; Karl Barth, *Church Dogmatics, Volume III: The Doctrine of Creation, Part 4*, trans. A. T. Mackay et al., ed. Geoffrey William Bromiley and Thomas F. Torrance (London: T&T Clark, 2004), 197–199.

36 Barth, *III/4*, 149–163.
37 RD2, 577; emphasis added.
38 Fuentes, *Why We Believe*, 27.
39 Ibid., 109.
40 Ibid., 40–41, 184–188; also see Agustín Fuentes, "Cooperation, Conflict and Niche Construction in the Genus *Homo*," in *War, Peace, and Human Nature: The Convergence of Cultural and Evolutionary Views*, ed. Douglas P. Fry (Oxford: Oxford University Press, 2013), 78–94.
41 Dietrich Bonhoeffer, *Sanctorum Communio: A Theological Study of the Sociology of the Church*, trans. Reinhard Krauss and Nancy Lukens, ed. Clifford J. Green and Joachim von Soosten, Dietrich Bonhoeffer Works, Volume 1 (Minneapolis: Fortress Press, 2009), 54–55.
42 Ibid., 56.
43 Shannon Vallor, *Technology and the Virtues: A Philosophical Guide to a Future Worth Wanting* (Oxford: Oxford University Press, 2016), 133.
44 Antonio R. Damasio, *Feeling and Knowing: Making Minds Conscious* (London: Robinson, 2021), 74–75.
45 Ibid., 28.
46 Ibid., 29.
47 It is beyond the scope of this study to examine empathy from the perspective of evolution theory; relevant studies on this, see Jean Decety, "The Neuroevolution of Empathy," *Annals of the New York Academy of Sciences* 1231, no. 1 (2011): 35–45.
48 Cristina Gonzalez-Liencres, Simone G. Shamay-Tsoory, and Martin Brüne, "Towards a Neuroscience of Empathy: Ontogeny, Phylogeny, Brain Mechanisms, Context and Psychopathology," *Neuroscience and Biobehavioral Reviews* 37 (2013): 1538.
49 Ibid., 1543–1544; also see Francis Stevens and Katherine Taber, "The Neuroscience of Empathy and Compassion in Pro-social Behavior," *Neuropsychologia* 159 (2021): 4–5.
50 See, e.g., Erin M. McDonald et al., "Activity Increase in Empathy-related Brain Regions When Children Contribute to Peers' Sadness and Happiness," *Journal of Experimental Child Psychology* 239 (2024): 105812; Frans B. M. de Waal and Stephanie D. Preston, "Mammalian Empathy: Behavioural Manifestations and Neural Basis," *Nature Reviews Neuroscience* 18 (2017): 506; J. Decety and K. J. Michalska, "A Neuroscience Perspective on Empathy and Its Development," in *Neural Circuit Development and Function in the Brain*, ed. John L. R. Rubenstein and Pasko Rakic (Oxford: Academic Press, 2013), 385–386; Claus Lamm, Jean Decety, and Tania Singer, "Meta-analytic Evidence for Common and Distinct Neural Networks Associated with Directly Experienced Pain and Empathy for Pain," *NeuroImage* 54 (2011): 2492–2502.
51 Stephanie D. Preston and Frans B. M. de Waal, "Empathy: Its Ultimate and Proximate Bases," *Behavioral and Brain Sciences* 25 (2002): 4.
52 See, e.g., Esslin L. Terrighena and Tatia M.C. Lee, "The Neuroimaging of Vicarious Pain," in *Neuroimaging of Pain*, ed. Luca Saba (Cham: Springer, 2017), 411–451. But it is worth noting that the mirror neuron mechanism does not blur the distinction between self and the other; see C. Chad Woodruff, "Reflections of Others and of Self: The Mirror Neuron System's Relationship to Empathy," in *The Neuroscience of Empathy, Compassion, and Self-Compassion*, ed. Larry Stevens and C. Chad Woodruff (London: Academic Press, 2018), 157–187.
53 Further on this, see Gregory R. Peterson, "Is My Feeling Your Pain Bad for Others? Empathy as Virtue versus Empathy as Fixed Trait," *Zygon: Journal of*

Religion & Science 52, no. 1 (2017): 232–257; Vallor, *Technology and the Virtues*, 132–138.

54 Joseph Weizenbaum, "ELIZA—A ComputerProgram For the Study of Natural Language Communication Between Man And Machine," *Communications of the ACM* 9, no. 1 (1966): 36–45.

55 Douglas Hofstadter, *Fluid Concepts and Creative Analogies: Computer Models of the Fundamental Mechanisms of Thought* (New York: Basic Books, 1995), 157. Further on ELIZA effects, see Sarah Dillon, "The Eliza Effect and Its Dangers: From Demystification to Gender Critique," *Journal for Cultural Research* 24, no. 1 (2020): 1–15.

56 Levy, "Falling in Love with a Companion," 92; emphasis added.

57 Turkle, *Alone Together*, 55.

58 Julia Knifka, "On the Significance of Understanding in Human–Robot Interaction," in *Social Robots: Boundaries, Potential, Challenges*, ed. Marco Nørskov (Farnham: Ashgate, 2016), 15.

59 Turkle, *Alone Together*, 107–108.

60 Paul Dumouchel and Luisa Damiano, *Living with Robots*, trans. Malcolm DeBevoise (Cambridge: Harvard University Press, 2017), 10.

61 Charles M. Ess, "What's Love Got to Do with It? Robots, Sexuality, and the Arts of Being Human," in *Social Robots: Boundaries, Potential, Challenges*, ed. Marco Nørskov (Farnham: Ashgate, 2016), 73.

62 Hashini Gunatilake and her colleagues flag up the underdeveloped research on empathy models in the field of software engineering. They suggest that software engineers should bring together different empathy models so as to enhance engineers' empathy for the needs of their end users. Hashini Gunatilake et al., "Empathy Models and Software Engineering—A Preliminary Analysis and Taxonomy," *The Journal of Systems and Software* 203 (2023): 11747.

63 Ramya Srinivasan and Beatriz San Miguel González, "The Role of Empathy for Artificial Intelligence Accountability," *Journal of Responsible Technology* 9 (2022): 5.

64 Shannon Vallor, "Carebots and Caregivers: Sustaining the Ethical Ideal of Care in the Twenty-First Century," *Philosophy & Technology* 24, no. 3 (2011): 251.

65 Ibid., 258–259.

66 Ibid., 258.

67 Aimee Van Wynsberghe, *Healthcare Robots: Ethics, Design and Implementation* (London: Routledge, 2016), 33.

68 Aimee Van Wynsberghe, "Care Ethics, Philosophy of Technology, and Robots in Humanitarian Action," in *The Oxford Handbook of Philosophy of Technology*, ed. Shannon Vallor (Oxford: Oxford University Press, 2021), 642; emphasis added.

69 This methodology is derived from yet different from the computer ethics approach 'value-sensitive design' (VSD), which 'is a theoretically grounded approach to the design of technology that accounts for human values in a principled and comprehensive manner throughout the design process;' Batya Friedman, Peter H. Kahn, Jr., and Alan Borning, "Value Sensitive Design and Information Systems," in *Human–Computer Interaction and Management Information Systems: Foundations*, ed. Ping Zhang and Dennis Galletta (Armonk, NY: M. E. Sharpe, 2006), 348–372. On the difference between CCVSD and VSD, see Van Wynsberghe, *Healthcare Robots*, 12–13.

70 Van Wynsberghe, *Healthcare Robots*, 82.

71 Ibid., 10.

72 Aimee Van Wynsberghe, "Designing Robots for Care: Care Centred Value-Sensitive Design," *Science and Engineering Ethics* 19, no. 2 (2013): 408.

73 Van Wynsberghe, *Healthcare Robots*, 70–76.
74 Ibid., 77.
75 David Lyall, *The Integrity of Pastoral Care* (London: SPCK, 2001), 12.
76 Young, "Virtual Pastor," 6–22.
77 Eric Stoddart, "Artificial Pastoral Care: Abdication, Delegation or Collaboration?," *Studies in Christian Ethics* 36, no. 3 (2023): 660–674. Hence, he argues: 'To abdicate care and leave it to algorithmic systems would be a mistake. To delegate care, however, is to draw on the gift of AI systems—but for us humans to retain our responsibilities for one another' (673).
78 Michael L. Littman et al., *Gathering Strength, Gathering Storms: The One Hundred Year Study on Artificial Intelligence (AI100) 2021 Study Panel Report*, Stanford University (Stanford, CA, September 2021), 68, http://ai100.stanford.edu/2021-report.

Conclusion

Let us circle back to the initial question: what has Jerusalem to do with Silicon Valley? This study has been dedicated to addressing this question through exploration of the connection yet radical distinction between AI and humans. The fundamental premise of this study is the precedence of ontology over ethics. From this it follows that the question about what AI *is* must be addressed before dealing with ethical issues surrounding various AI applications.

The theology of the *imago Dei* and the thinking of archetype–ectype together give shape to a theo-ontological foundation for AI ethics: AI as the *imago hominis* is the ectype of the *imago Dei*, which means AI is radically distinct from yet closely connected to the human being as the archetype. This theo-ontology is tied up with a holistic reading of the *imago Dei*—that is, both the human body and the mind are part of the image of God, which is the signifier of the beginning of both God–human stories and stories of human ethical performances towards others. This psychosomatic nature of humanity features in the uniqueness of human beings compared to AI. Both the human body and the mind play a part in human performances towards God and others. Seen through the lens of the ontological distinction between AI and humans, the connection between humans and AI can be aptly construed: AI as *imago hominis* is expected to participate in God–human stories and stories of human ethical performances towards others, without the attempt to design AI as the substitute for humans in religious and ethical performances.

This radical distinction yet connection between humans and AI has become evident after looking into the concept of artificial consciousness. Despite the bold claim made by some scholars that the artificial neural networks can be used to reproduce human consciousness in AI systems, both the biological substrate of the mind and human biological particularities cannot be reduced to mathematical models or AI algorithms. It is in the sense of simulation, rather than emulation, that artificial consciousness should be understood in relation to the creature consciousness of human beings. In ontological terms, AI's simulated consciousness is ectypal, which means that

DOI: 10.4324/9781003356738-7

AI has the potential to mirror human religious and moral consciousness. Coupled with the particularity of human biological conditions, this distinction yet connection between human and artificial consciousness can be illustrated through exploration of AI-powered liturgical enactment, AI's moral agency, and AI-assisted companionship.

My comparative analysis of human and AI-powered liturgical enactment has elucidated that both human bodies and minds play a unique role in liturgical enactment vis-à-vis AI ministers. The detailed treatment of the bodily action, the mirror neuron mechanism, and the memory of human beings showcases that psychosomatic humanity cannot be replaced by AI altogether in authentic liturgical enactment. That said, AI-powered liturgical enactment can be viewed as augmenting liturgies and extending the invitation for the liturgy broadly from the perspective of AI as the ectype of humanity.

The exploration of the concept of artificial moral agency has led to a similar corollary regarding ethics. It is misleading to consider AI as a morally neutral instrument and belittle its impacts on human moral life. Yet, it is also an exaggeration to claim that AI possesses the same full moral agency as human agents precisely because AI cannot emulate the close tie between the human body, the mind, and morality. The theo-ontology of AI gives birth to the concept of ectypal partial moral agency of AI, which highlights both AI's bearing on human moral life and the responsibility *connection* between human agents and AI. In this vein, the theo-ontology of AI shows that AI as the ectype of humanity participates in the ethical performance between humans and fosters the collective well-being of the community of the *imago Dei*.

The theo-ontology of AI has strengths in conceptualising interaction between humans and AI, particularly in exploring AI-assisted companionship. Rather than employing the substantive, functional, or relational approaches to artificial companionship, I have leveraged the idea of AI as the ectype of *imago Dei* to craft the concept of companionable AI as the ectype of human companions by virtue of human psychosomatic nature. The neural activities, feelings, empathy, and other bodily and mental actions of human beings showcase the unique nature of human companionship compared to AI-assisted companionship. Hence, 'companionable AI' is an apparatus through which to bring to light AI's augmentation of the human capacity for companionship.

That being so, this study has demonstrated that the hype surrounding AI like AGI and ASI results from the ill-grounded imaginations of AI or unawareness of the ontological distinction between humans and AI. As Daniel Dennett remarks, '[t]he gap between today's systems and the science-fictional systems dominating the popular imagination is still huge, though many folks, both lay and expert, manage to underestimate it.'[1] The theo-ontology of AI constructed in this study offers antidotes to these imaginations by unpacking the ontology of the psychosomatic human being vis-à-vis AI. By doing so, it has successfully shifted the centre of gravity from fictional AI to real-life AI that is pervasive in human lives, from speculations about AI to AI ethics.[2]

AI ethics hold great significance for future AI–theology dialogues. With the increasingly wide range of AI applications, the hype surrounding AI will 'deflate significantly,' as Mark Coeckelbergh predicts.[3] For AI will be fully integrated into human lives and be demystified with the populace's growing knowledge of AI. If AI will become as ordinary as other technologies, AI–theology research should pay more attention to what theological ethics can offer to address moral, societal, religious issues surrounding AI applications, rather than to speculations about AI like AGI or ASI, which are dominant in the present literature on AI and religion.

From the vantage point of the theo-ontology of AI, it can be argued that theology and AI research can be mutually enhanced. On the one hand, theologians can become a member of 'many hands' who are involved in the design and creation of AI, bringing together theological tools to address ethical challenges posed by AI and contributing towards a moral future of AI. On the other hand, AI researchers and designers can help theologians to settle into the age of AI by opening the black box of AI technologies, which can orient theological reflections on AI as well as other digital technologies. This will expand theology's reach to address the most recent challenges posed by AI to religious communities. In the end, the divide between Jerusalem and Silicon Valley will be bridged, and AI will be properly understood as the *imago hominis*, the ectype of the *imago Dei*.

Notes

1 Daniel C. Dennett, "What Can We Do?," in *Possible Minds: 25 Ways of Looking at AI*, ed. John Brockman (New York: Penguin Books, 2019), 49.

2 A similar emphasis on AI ethics from a Christian theological perspective can be found in George Coghill, "Artificial Intelligence (and Christianity): Who? What? Where? When? Why? and How?," *Studies in Christian Ethics* 36, no. 3 (2023): 604–619; Michael Burdett, "Proximate and Ultimate Concerns in Christian Ethical Responses to Artificial Intelligence," *Studies in Christian Ethics* 36, no. 3 (2023): 620–641.

3 Coeckelbergh, *AI Ethics*, 81.

Bibliography

Aquinas, Thomas. *Summa Theologiae*. Translated by Fr. Laurence Shapcote. Edited by The Aquinas Institute. Latin/English Edition of the Works of St. Thomas Aquinas, Vols. 13–20. Green Bay: Aquinas Institute, 2012.

Atkins, Peter. *Memory and Liturgy: The Place of Memory in the Composition and Practice of Liturgy*. Aldershot: Ashgate, 2004.

Atkinson, Richard C., and Richard M. Shiffrin. "The Control of Short-Term Memory." *Scientific American* 225, no. 2 (1971): 82–91.

Atkinson, Richard C., and Richard M. Shiffrin. "Human Memory: A Proposed System and Its Control Processes." In *The Psychology of Learning and Motivation*. Edited by K. Spence and J. Spence, 89–105. New York: Academic Press, 1968.

Augustine. *Concerning the City of God against the Pagans*. Translated by Henry Bettenson. London: Penguin Books, 2003.

Augustine. *Confessions*. Translated by Henry Chadwick. Oxford: Oxford University Press, 1998.

Augustine. "Marriage and Desire." In *Answer to the Pelagians, II*. Edited by John E. Rotelle. Translated by Roland Teske. Works of Saint Augustine, I/24. 12–97. New York: New City Press, 1998.

Austin, Victor Lee. *Christian Ethics: A Guide for the Perplexed*. London: Bloomsbury, 2012.

Baddeley, Alan D. *Essentials of Human Memory*. East Sussex: Psychology Press, 1999.

Balle, Simon, and Ulrik Nissen. "Responsive Bodies: Robots, AI, and the Question of Human Distinctiveness." *Zygon: Journal of Religion & Science* 58, no. 2 (2023): 358–377.

Barbour, Ian G. "Neuroscience, Artificial Intelligence, and Human Nature: Theological and Philosophical Reflection." *Zygon: Journal of Religion and Science* 34, no. 3 (1999): 361–398.

Barth, Karl. *Church Dogmatics, Volume III: The Doctrine of Creation, Part 1*. Translated by J. W. Edwards, O. Bussey and H. Knight. Edited by Geoffrey William Bromiley and Thomas F. Torrance. London: T&T Clark, 2004.

Barth, Karl. *Church Dogmatics, Volume III: The Doctrine of Creation, Part 4*. Translated by A. T. Mackay, T. H. L. Parker, H. Knight, H. A. Kennedy and J. Marks. Edited by Geoffrey William Bromiley and Thomas F. Torrance. London: T&T Clark, 2004.

Barth, Karl. *Die kirchliche Dogmatik, Band III,1*. Zürich: Theologischer Verlag Zürich, 1980.

Bavinck, Herman. *The Philosophy of Revelation: The Stone Lectures for 1908-1909*. New York: Longmans, Green, and Co, 1909.

Bavinck, Herman. *Gereformeerde Dogmatiek, Tweede Deel*. 4th ed. Kampen: J. H. Kok, 1928.

Bavinck, Herman. *Reformed Dogmatics, Volume 1: Prolegomena.* Translated by John Vriend. Edited by John Bolt. Grand Rapids: Baker, 2003.

Bavinck, Herman. *Reformed Dogmatics, Volume 2: God and Creation.* Translated by John Vriend. Edited by John Bolt. Grand Rapids: Baker, 2004.

Bavinck, Herman. "Evolution." In *Essays on Religion, Science, and Society.* Edited by Bolt, John. Translated by Harry Boonstra and Gerrit Sheeres. 105–118. Grand Rapids: Baker, 2008.

Bavinck, Herman. *The Christian Family.* Translated by Nelson Kloosterman. Grand Rapids: Christian's Library Press, 2012.

Bavinck, Herman. *Reformed Ethics, Volume I: Created, Fallen, and Converted Humanity.* Edited by John Bolt, Jessica Joustra, Nelson D. Kloosterman, Antoine Theron and Dirk Van Keulen. Grand Rapids: Baker Academic, 2019.

Bavinck, Herman. *Reformed Ethics, Volume II: The Duties of the Christian Life.* Edited by John Bolt, Jessica Joustra, Nelson Kloosterman, Antoine Theron and Dirk Van Keulen. Grand Rapids, Michigan: Baker Academic, 2021.

Belcher, Kimberly H., Kevin G. Grove, and Sonja K. Pilz. "Recording as the Re-Membering Work of the People: A Catholic-Jewish Dialogue on the Body and Liturgical Memory." *Studia Liturgica* 51, no. 2 (2021): 122–142.

Berger, Teresa. *@ Worship: Liturgical Practices in Digital Worlds.* London: Routledge, 2018.

Berry, R. J. "Biology Since Darwin." In *Darwinism and Natural Theology.* Edited by Andrew Robinson, 12–38. Newcastle: Cambridge Scholars, 2012.

Block, Ned. "On a Confusion about a Function of Consciousness." *Behavioral and Brain Sciences* 18 (1995): 227–247.

Boden, Margaret A. *AI: Its Nature and Future.* Oxford: Oxford University Press, 2016.

Boden, Margaret A. *Artificial Intelligence: A Very Short Introduction.* Oxford: Oxford University Press, 2018.

Bonhoeffer, Dietrich. *Sanctorum Communio: A Theological Study of the Sociology of the Church.* Translated by Reinhard Krauss and Nancy Lukens. Edited by Clifford J. Green and Joachim von Soosten. Dietrich Bonhoeffer Works, Volume 1. Minneapolis: Fortress Press, 2009.

Braaten, Carl E., and LaVonne Braaten. *The Living Temple: A Practical Theology of the Body and the Foods of the Earth.* Eugene, Oregon: Wipf & Stock, 1976.

Brey, Philip. "The Social Ontology of Virtual Environments." *American Journal of Economics and Sociology* 62, no. 1 (2003): 269–282.

Brock, Cory. *Orthodox yet Modern: Herman Bavinck's Use of Friedrich Schleiermacher.* Bellingham: Lexham, 2020.

Brooke, John Hedley. "Visions of Perfectibility." *Journal of Evolution and Technology* 14, no. 2 (2005): 1–12.

Brown, Warren S. "Nonreductive Physicalism and Soul: Finding Resonance Between Theology and Neuroscience." *American Behavioral Scientist* 45, no. 12 (2002): 1812–1821.

Bryson, Joanna J. "Robots Should Be Slaves." In *Close Engagements with Artificial Companions: Key Social, Psychological, Ethical and Design Issues.* Edited by Yorick Wilks. Amsterdam: John Benjamins, 2010.

Bryson, Joanna J. "The Artificial Intelligence of the Ethics of Artificial Intelligence: An Introductory Overview for Law and Regulation." In *The Oxford Handbook of Ethics of AI.* Edited by Markus D. Dubber, Frank Pasquale, and Sunit Das, 3–25. Oxford: Oxford University Press, 2020.

Burdett, Michael. *Eschatology and the Technological Future.* New York: Routledge, 2015.

Burdett, Michael. "Proximate and Ultimate Concerns in Christian Ethical Responses to Artificial Intelligence." *Studies in Christian Ethics* 36, no. 3 (2023): 620–641.

Calvin, John. *Institutes of the Christian Religion, Volume 1 & 2*. Translated by Ford Lewis Battles. Edited by John T. McNeill. Louisville, Kentucky: Westminster John Knox Press, 2011.

Cellan-Jones, Rory. "Stephen Hawking Warns Artificial Intelligence Could End Mankind." *BBC News*, November 7, 2017. https://www.bbc.co.uk/news/technology-30290540.

Chalmers, David. "Facing Up to the Problem of Consciousness." In *Explaining Consciousness—The "Hard Problem"*. Edited by Jonathan Shear, 9–30. Cambridge, Massachusetts: MIT Press, 1997.

Chalmers, David. "The Hard Problem of Consciousness." In *The Blackwell Companion to Consciousness*. Edited by Susan Schneider and Max Velmans, 32–42. Chichester: Wiley-Blackwell, 2017.

Chella, Antonio, Marcello Frixione, and Salvatore Gaglio. "A Cognitive Architecture for Robot Self-Consciousness." *Artificial Intelligence in Medicine* 44 (2008): 147–154.

Chin, Chuanfei. "Artificial Consciousness: From Impossibility to Multiplicity." In *Philosophy and Theory of Artificial Intelligence 2017*. Edited by C. Müller. Vincent, 3–18. Cham: Springer, 2018.

Chou, Wen-Chuang, and Tsung-Ying Sun. "The Computational Model to Simulate the Progress of Perceiving Patterns in Neuron Population." In *Artificial Neural Networks: Biological Inspirations – ICANN 2005*. Edited by Włodzisław Duch, Janusz Kacprzyk, Erkki Oja and Sławomir Zadrożny, 7–12. Berlin: Springer, 2005.

Chow, Alexander, and Jonas Kurlberg. "Two or Three Gathered Online: Asian and European Responses to COVID-19 and the Digital Church." *Studies in World Christianity* 26, no. 3 (2020): 298–318.

Chrisley, Ron. "Philosophical Foundations of Artificial Consciousness." *Artificial Intelligence in Medicine* 44 (2008): 119–137.

Churchland, Patricia S. *Braintrust: What Neuroscience Tells Us About Morality*. Princeton: Princeton University Press, 2011.

Clowes, Robert W., and Anil K. Seth. "Axioms, Properties and Criteria: Roles for Synthesis in the Science of Consciousness." *Artificial Intelligence in Medicine* 44 (2008): 91–104.

Cockayne, Joshua, and Gideon Salter. "Feasts of Memory: Collective Remembering, Liturgical Time Travel and the Actualisation of the Past." *Modern Theology* 37, no. 2 (2021): 275–295.

Coeckelbergh, Mark. "Virtual Moral Agency, Virtual Moral Responsibility: On the Moral Significance of the Appearance, Perception, and Performance of Artificial Agents." *AI & Society* 24 (2009): 181–189.

Coeckelbergh, Mark. *AI Ethics*. Cambridge, Massachusetts: The MIT Press, 2020.

Coeckelbergh, Mark. "Artificial Intelligence, Responsibility Attribution, and a Relational Justification of Explainability." *Science and Engineering Ethics* 26 (2020): 2051–2068. https://doi-org.ezproxy.is.ed.ac.uk/10.1007/s11948-019-00146-8.

Coghill, George. "Artificial Intelligence (and Christianity): Who? What? Where? When? Why? and How?." *Studies in Christian Ethics* 36, no. 3 (2023): 604–619.

Coleman, Kari Gwen. "Android Arete: Toward a Virtue Ethic for Computational Agents." *Ethics and Information Technology* 3 (2001): 247–265.

Constantinescu, Mihaela, and Roger Crisp. "Can Robotic AI Systems Be Virtuous and Why Does This Matter?" *International Journal of Social Robotics* 14 (2022): 1547–1557.

Cortez, Marc. *Embodied Souls, Ensouled Bodies: An Exercise in Christological Anthropology and Its Significance for the Mind/Body Debate*. London: T&T Clark, 2008.

Crawford, Kate. *Atlas of AI: Power, Politics, and the Planetary Costs of Artificial Intelligence*. New Haven: Yale University Press, 2021.

Crichton, J. D. "A Theology of Worship." In *The Study of Liturgy*. Edited by Cheslyn Jones, Geoffrey Wainwright and Edward Yarnold, 1–29. London: SPCK, 1978.

Crisp, Oliver. "A Christological Model of the *Imago Dei*." In *The Ashgate Research Companion to Theological Anthropology*. Edited by Joshua Ryan Farris and Charles Taliaferro, 217–229. Farnham: Ashgate, 2015.

Curry, Eugene A. "Artificial Intelligence and Baptism: Cutting a Gordian Knot." *Theology and Science* 20, no. 2 (2022): 156–165.

Damasio, Antonio R. "Neuroscience and Ethics: Intersections." *The American Journal of Bioethics* 7, no. 1 (2007): 3–7.

Damasio, Antonio R. *Feeling and Knowing: Making Minds Conscious*. London: Robinson, 2021.

Damasio, Antonio R. *The Strange Order of Things: Life, Feeling and the Making of Cultures*. London: Robinson, 2021.

Daston, Lorraine. *Rules: A Short History of What We Live By*. Princeton: Princeton University Press, 2022.

Dawn, Marva J. *A Royal "Waste" of Time: The Splendor of Worshiping God and Being Church for the World*. Grand Rapids: Eerdmans, 1999.

de Waal, Frans B. M., and Stephanie D. Preston. "Mammalian Empathy: Behavioural Manifestations and Neural Basis." *Nature Reviews Neuroscience* 18 (2017): 498–509.

Deane-Drummond, Celia. "The Myth of Moral Bio-Enhancement: An Evolutionary Anthropology and Theological Critique." In *Religion and Human Enhancement: Death, Values, and Morality*. Edited by Tracy J. Trothen and Calvin Mercer, 175–190. Cham: Palgrave Macmillan, 2017.

Deane-Drummond, Celia E. "God's Image and Likeness in Humans and Other Animals: Performative Soul-Making and Graced Nature." *Zygon: Journal of Religion and Science* 47, no. 4 (2012): 934–948.

Deane-Drummond, Celia E. "Response: *Homo Divinus*–Myth or Reality?" In *Darwinism and Natural Theology: Evolving Perspectives*. Edited by Andrew Robinson, 39–46. Cambridge: Cambridge Scholars, 2012.

Deane-Drummond, Celia E. "In God's Image and Likeness: From Reason to Revelation in Humans and Other Animals." In *Questioning the Human: Toward a Theological Anthropology for the Twenty-First Century*. Edited by Lieven Boeve, Yves de Maeseneer, and Ellen Van Stichel, 60–76. Oxford: Oxford University Press, 2014.

Deane-Drummond, Celia E. *The Wisdom of the Liminal: Evolution and Other Animals in Human Becoming*. Grand Rapids: Eerdmans, 2014.

Deane-Drummond, Celia E., Neil Arner, and Agustín Fuentes. "The Evolution of Morality: A Three-Dimensional Map." *Philosophy, Theology and the Sciences* 3, no. 2 (2016): 115–151.

Deane-Drummond, Celia E. *Theological Ethics through a Multispecies Lens*. The Evolution of Wisdom, Volume I. Oxford: Oxford University Press, 2019.

Decety, J., and K. J. Michalska. "A Neuroscience Perspective on Empathy and Its Development." In *Neural Circuit Development and Function in the Brain*. Edited by John L. R. Rubenstein and Pasko Rakic, 379–393. Oxford: Academic Press, 2013.

Decety, Jean. "The Neuroevolution of Empathy." *Annals of the New York Academy of Sciences* 1231, no. 1 (2011): 35–45.

Delio, Ilia. *Re-Enchanting the Earth: Why AI Needs Religion*. Maryknoll: Orbis Books, 2020.

Dennett, Daniel C. "What Can We Do?." In *Possible Minds: 25 Ways of Looking at AI*. Edited by John Brockman, 41–53. New York: Penguin Books, 2019.

Dennett, Daniel C., and Marcel Kinsbourne. "Time and the Observer: The Where and When of Consciousness in the Brain." *Behavioral and Brain Sciences* 15 (1992): 183–247.

Dignum, Virginia. *Responsible Artificial Intelligence: How to Develop and Use AI in a Responsible Way*. Cham: Springer, 2019.

Dillon, Sarah. "The Eliza Effect and Its Dangers: From Demystification to Gender Critique." *Journal for Cultural Research* 24, no. 1 (2020): 1–15.

Dodig-Crnkovic, Gordana. "The Info-Computational Nature of Morphological Computing." In *Philosophy and Theory of Artificial Intelligence*. Edited by Vincent C. Müller, 59–68. Berlin: Springer, 2013.

Dreyfus, Hubert L. *What Computers Can't Do: A Critique of Artificial Reason*. New York: Harper & Row, 1972.

Dreyfus, Hubert L. *What Computers Still Can't Do: A Critique of Artificial Reason*. Revised ed. Cambridge, Massachusetts: MIT Press, 1992.

Dumouchel, Paul, and Luisa Damiano. *Living with Robots*. Translated by Malcolm DeBevoise. Cambridge, Massachusetts: Harvard University Press, 2017.

Durante, Massimo. "Technology and the Ontology of the Virtual." In *Oxford Handbook of Philosophy of Technology*. Edited by Shannon Vallor, 318–340. Oxford: Oxford University Press, 2022.

Dyrness, William A. *A Primer on Christian Worship: Where We've Been, Where We Are, Where We Can Go*. Grand Rapids: Eerdmans, 2009.

Eglinton, James. *Trinity and Organism: Towards a New Reading of Herman Bavinck's Organic Motif*. London: T&T Clark, 2012.

Ellul, Jacques. *The New Demons*. Translated by C. Edward Hopkin. New York: Seabury, 1975.

Ess, Charles M. "What's Love Got to Do with It? Robots, Sexuality, and the Arts of Being Human." In *Social Robots: Boundaries, Potential, Challenges*. Edited by Marco Nørskov, 57–79. Farnham: Ashgate, 2016.

Farwell, James. *This Is the Night: Suffering, Salvation, and the Liturgies of Holy Week*. London: T&T Clark, 2005.

Fergusson, David. "Are We Alone? And Does It Matter?: The Narrative of Human Particularity." In *Human Origins and the Image of God: Essaysin Honor of J. Wentzel van Huyssteen*. Edited by Christopher Lilley and Daniel J. Pedersen, 236–249. Grand Rapids: Eerdmans, 2017.

Fergusson, David. *Creation*. Grand Rapids: Eerdmans, 2014.

Fergusson, David. "Humans Created According to the *Imago Dei*: An Alternative Proposal." *Zygon: Journal of Religion and Science* 48, no. 2 (2013): 439–453.

Flach, Peter. *Machine Learning: The Art and Science of Algorithms that Make Sense of Data*. Cambridge: Cambridge University Press, 2012.

Floridi, Luciano. "Artificial Companions and Their Philosophical Challenges." In *Close Engagements with Artificial Companions: Key Social, Psychological, Ethical and Design Issues*. Edited by Yorick Wilks. Amsterdam: John Benjamins, 2010.

Floridi, Luciano. "Artificial Intelligence's New Frontier: Artificial Companions and the Fourth Revolution." *Metaphilosophy* 39, no. 4/5 (2008): 651–655.

Floridi, Luciano. *The Ethics of Information*. Oxford: Oxford University Press, 2013.

Floridi, Luciano. *The Fourth Revolution: How the Infosphere Is Reshaping Human Reality*. Oxford: Oxford University Press, 2014.

Floridi, Luciano. "Information Ethics." In *The Cambridge Handbook of Information and Computer Ethics*. Edited by Luciano Floridi, 77–98. Cambridge: Cambridge University Press, 2010.

Floridi, Luciano. "On the Morality of Artificial Agents." In *Machine Ethics*. Edited by Michael Anderson and Susan Leigh Anderson, 184–212. Cambridge: Cambridge University Press, 2011.

Floridi, Luciano. *The Philosophy of Information*. Oxford: Oxford University Press, 2011.

Floridi, Luciano, and J. W. Sanders. "Artificial Evil and the Foundation of Computer Ethics." *Ethics and Information Technology* 3, no. 1 (2001): 55–66.

Floridi, Luciano, and J. W. Sanders. "On the Morality of Artificial Agents." *Minds and Machine* 14, no. 3 (2004): 349–379.

Floridi, Luciano, and J. W. Sanders. "The Method of Abstraction." In *Yearbook of the Artificial: Nature, Culture and Technology. Vol. 2: Models in Contemporary Sciences*. Edited by Massimo Negrotti, 177–220. Bern: Peter Lang, 2004.

Foerst, Anne. *God in the Machine: What Robots Teach Us about Humanity and God*. New York: Plume, 2005.

Foresight Mental Capital and Wellbeing. *Mental Capital and Wellbeing: Making the Most of Ourselves in the 21st Century*. London: The Government Office for Science, 2008.

Francis, Pope. "The Common Good in the Digital Age." (27 September 2019). https://www.vatican.va/content/francesco/en/speeches/2019/september/documents/papa-francesco_20190927_eradigitale.html.

Franklin, Stan. "IDA: A Conscious Artifact?." *Journal of Consciousness Studies* 10, no. 4–5 (2003): 47–66.

Friedman, Batya, Peter H. Kahn Jr., and Alan Borning. "Value Sensitive Design and Information Systems." In *Human–Computer Interaction and Management Information Systems: Foundations*. Edited by Ping Zhang and Dennis Galletta, 348–372. Armonk: M. E. Sharpe, 2006.

Frith, Chris D., and Geraint Rees. "A Brief History of the Scientific Approach to the Study of Consciousness." In *The Blackwell Companion to Consciousness*. Edited by Susan Schneider and Max Velmans, 3–16. Chichester: Wiley-Blackwell, 2017.

Fuentes, Agustín. "Cooperation, Conflict and Niche Construction in the Genus *Homo*." In *War, Peace, and Human Nature: The Convergence of Cultural and Evolutionary Views*. Edited by Douglas P. Fry, 78–94. Oxford: Oxford University Press, 2013.

Fuentes, Agustín. *Why We Believe: Evolution and the Human Way of Being*. New Haven: Yale University Press, 2019.

Fuller, Robert C., and Derek E. Montgomery. "Body Posture and Religious Attitudes." *Archive for the Psychology of Religion* 37 (2015): 227–239.

Gay, Doug, and Ron Rienstra. "Veering Off the Via Media: Emerging Church, Alternative Worship, and New Media Technologies in the United States and United Kingdom." *Liturgy* 23, no. 3 (2008): 39–47.

Geldhof, Joris. "Penetration–Permeation–Fermentation: Ponderings on the Being of Liturgy and Its Memorial Modes." *Studia Liturgica* 50, no. 1 (2020): 6–21.

Geraci, Robert M. "Spiritual Robots: Religion and Our Scientific View of the Natural World." *Theology and Science* 4, no. 3 (2006): 229–246.

Geraci, Robert M. "Robots and the Sacred in Science and Science Fiction: Theological Implications of Artificial Intelligence." *Zygon: Journal of Religion and Science* 42, no. 4 (2007): 961–980.

Geraci, Robert M. "Apocalyptic AI: Religion and the Promise of Artificial Intelligence." *Journal of the American Academy of Religion* 76, no. 1 (2008): 138–166.

Geraci, Robert M. *Apocalyptic AI: Visions of Heaven in Robotics, Artificial Intelligence, and Virtual Reality*. Oxford: Oxford University Press, 2010.

Gerhard, Johann. *On the Nature of Theology and on Scripture*. Translated by Richard J. Dinda. Edited by Benjamin T. G. Mayes. Saint Louis: Concordia Publishing House, 2009.

Ghiglino, Davide, and Agnieszka Wykowska. "When Robots (Pretend to) Think." In *Artificial Intelligence: Reflections in Philosophy, Theology, and the Social Sciences*. Edited by Benedikt Paul Göcke and Astrid Rosenthal-von der Pütten, 49–74. Paderborn: Mentis, 2020.

Gingles, Dallas. "Narrative." In *T&T Clark Handbook of Christian Ethics*. Edited by Tobias Winright, 103–110. London: T&T Clark, 2021.

Gittoes, Julie. *Anamnesis and the Eucharist: Contemporary Anglican Approaches*. Aldershot: Ashgate, 2008.

Goel, Ashok K., and Jim Davies. "Artificial Intelligence." In *The Cambridge Handbook of Intelligence*. Edited by Robert J. Sternberg and Scott Barry Kaufman, 468–482. Cambridge: Cambridge University Press, 2011.

Gonzalez-Liencres, Cristina, Simone G. Shamay-Tsoory, and Martin Brüne. "Towards a Neuroscience of Empathy: Ontogeny, Phylogeny, Brain Mechanisms, Context and Psychopathology." *Neuroscience and Biobehavioral Reviews* 37 (2013): 1537–1548.

Gordon, John-Stewart. "Building Moral Robots: Ethical Pitfalls and Challenges." *Science and Engineering Ethics* 26 (2020): 141–157.

Graham, George, Terence Horgan, and John Tienson. "Consciousness and Intentionality." In *The Blackwell Companion to Consciousness*. Edited by Susan Schneider and Max Velmans, 519–535. Chichester: Wiley-Blackwell, 2017.

Grenz, Stanley J. *The Social God and the Relational Self: A Trinitarian Theology of the Imago Dei*. Louisville: Westminster John Knox Press, 2001.

Grieshaber, Kirstern. "Can a Chatbot Preach a Good Sermon? Hundreds Attend Church Service Generated by Chatgpt to Find Out." *AP News*, 10 June 2023. https://apnews.com/article/germany-church-protestants-chatgpt-ai-sermon-651f21c24cfb47e3122e987a7263d348.

Grodzinsky, Frances S., Keith W. Miller, and Marty J. Wolf. "The Ethics of Designing Artificial Agents." *Ethics and Information Technology* 10 (2008): 115–121.

Gunatilake, Hashini, John Grundy, Ingo Mueller, and Rashina Hoda. "Empathy Models and Software Engineering—A Preliminary Analysis and Taxonomy." *The Journal of Systems and Software* 203 (2023): 11747.

Gunkel, David J. *The Machine Question: Critical Perspectives on AI, Robots, and Ethics*. Cambridge, Massachusetts: MIT, 2012.

Hakli, Raul, and Pekka Mäkelä. "Moral Responsibility of Robots and Hybrid Agents." *The Monist* 102 (2019): 259–275.

Haksar, Vinit. "Moral Agents." In *Routledge Encyclopedia of Philosophy*. Edited by Edward Craig. London: Routledge, 1998. https://doi.org/10.4324/9780415249126-L049-1.

Hanson, F. Allan. "Beyong the Skin Bag: On the Moral Responsibility of Extended Agencies." *Ethics and Information Technology* 11 (2009): 91–99.

Harari, Yuval Noah. *Homo Deus: A Brief History of Tomorrow*. London: Vintage, 2017.

Hardalupas, Mahi. "A Systematic Account of Machine Moral Agency." In *Philosophy and Theory of Artificial Intelligence 2017*. Edited by Vincent C. Müller, 252–254. Cham: Springer, 2018.

Harnad, Stevan, and Peter Scherzer. "First, Scale up to the Robotic Turing Test, Then Worry about Feeling." *Artificial Intelligence in Medicine* 44 (2008): 83–89.

Harrison, N. V. "Introduction." In *Gregory of Nazianzus, Festal Orations*. 11–56. Crestwood: St. Vladimir's Seminary Press, 2008.

Hart, David B., and John Chryssavgis, eds. *For the Life of the World: Toward a Social Ethos of the Orthodox Church*. Brookline: Holy Cross Orthodox Press, 2020.

Heaven, Will Douglas. "Meta's New Learning Algorithm Can Teach AI to Multi-Task." *MIT Technology Review*, 20 January 2022. https://www.technologyreview.com/2022/01/20/1043885/meta-ai-facebook-learning-algorithm-nlp-vision-speech-agi/.

Hefner, Philip. "The Church." In *Christian Dogmatics, Volume 1 & 2*. Edited by Carl E. Braaten and Robert W. Jenson. 2:179–247. Philadelphia: Fortress, 2011.
Hefner, Philip. *The Human Factor: Evolution, Culture, and Religion*. Minneapolis: Fortress, 1993.
Hegel, G. W. F. "The Positivity of the Christian Religion." In *On Christianity: Early Theological Writings* Translated by T. M. Knox, 67–181. New York: Harper & Brothers, 1961.
Heidegger, Martin. *The Question Concerning Technology and Other Essays*. Translated by William Lovitt. New York: Garland, 1977.
Herzfeld, Noreen. "Creating in Our Own Image: Artificial Intelligence and the Image of God." *Zygon: Journal of Religion and Science* 37, no. 2 (2002): 303–316.
Herzfeld, Noreen. *In Our Image: Artificial Intelligence and the Human Spirit*. Minneapolis: Fortress, 2002.
Herzfeld, Noreen. "Co-*Creator* or *Co*-Creator?: The Problem with Artificial Intelligence." In *Creative Creatures: Values and Ethical Issues in Theology, Science and Technology*. Edited by Ulf Görman, Willem Drees, and Hubert Meisinger. London: T&T Clark, 2005.
Herzfeld, Noreen. "Bodies Matter: A New Fad and a Fallacy in the Name of Science." In *God's Action in Nature's World: Essays in Honour of Robert John Russell*. Edited by Ted Peters and Nathan Hallanger, 225–233. Aldershot: Ashgate, 2006.
Herzfeld, Noreen. "A New Member of the Family? The Continuum of Being, Artificial Intelligence, and the Image of God." *Theology and Science* 5, no. 3 (2007): 235–247.
Herzfeld, Noreen. *Technology and Religion: Remaining Human in a Co-created World*. West Conshohocken: Templeton Press, 2009.
Herzfeld, Noreen. "In Whose Image? Artificial Intelligence and the Imago Dei." In *The Blackwell Companion to Science and Christianity*. Edited by J. B. Stump, and Alan G. Padgett, 500–509. Chichester: Wiley-Blackwell, 2012.
Herzfeld, Noreen. *The Artifice of Intelligence: Divine and Human Relationship in a Robotic Age*. Minneapolis: Fortress, 2023.
Herzog, Christian. "Three Risks That Caution Against a Premature Implementation of Artificial Moral Agents for Practical and Economical Use." *Science and Engineering Ethics* 27 (2021): 3.
Himma, Kenneth Einar. "Artificial Agency, Consciousness, and the Criteria for Moral Agency: What Properties Must an Artificial Agent Have to Be a Moral Agent?" *Ethics and Information Technology* 11 (2009): 19–29.
Hofstadter, Douglas. *Fluid Concepts and Creative Analogies: Computer Models of the Fundamental Mechanisms of Thought*. New York: Basic Books, 1995.
Hogue, David A. "Sensing the Other in Worship: Mirror Neurons and the Empathizing Brain." *Liturgy* 21, no. 3 (2006): 31–39.
Holland, Owen, and D. Gamez. "Artificial Intelligence and Consciousness." In *Encyclopedia of Consciousness*. Edited by W. P. Banks, 37–45. Oxford: Academic, 2009.
Holland, Owen, Rob Knight, and Richard Newcombe. "A Robot-Based Approach to Machine Consciousness." In *Artificial Consciousness*. Edited by Antonio Chella and Riccardo Manzotti, 156–173. Exeter: Imprint Academic, 2007.
Hunter, David G. "Augustine on the Body." In *A Companion to Augustine*. Edited by Mark Vessey and Shelley Reid, 353–364. Chichester: Wiley-Blackwell, 2012.
Hurtado, Larry W. *At the Origins of Christian Worship: The Context and Character of Earliest Christian Devotion*. Grand Rapids: Eerdmans, 1999.
Hutter, Marcus. "One Decade of Universal Artificial Intelligence." In *Theoretical Foundations of Artificial General Intelligence*. Edited by Pei Wang and Ben Goertzel, 67–88. Paris: Atlantis, 2012.
Jackelén, Antje. *Time and Eternity: The Question of Time in Church, Science, and Theology*. Philadelphia: Templeton Foundation Press, 2005.

Jeeves, Malcolm. "Brains, Minds, Souls, and People: A Scientific Perspective on Complex Human Personhood." In *The Depth of the Human Person: A Multidisciplinary Approach*. Edited by Michael Welker, 93–108. Grand Rapids: Eerdmans, 2014.

Johnson, Deborah G. "Computer Systems: Moral Entities but Not Moral Agents." *Ethics and Information Technology* 8 (2006): 195–204.

Johnson, Deborah G. and Keith W. Miller. "Un-making Artificial Moral Agents." *Ethics and Information Technology* 10 (2008): 123–133.

Johnson, Deborah G. and Thomas M. Powers. "Computers as Surrogate Agents." In *Information Technology and Moral Philosophy*. Edited by Jeroen van den Hoven and John Weckert, 251–269. Cambridge: Cambridge University Press, 2008.

Johnson, Mark. *The Body in the Mind: The Bodily Basis of Meaning, Imagination, and Reason*. Chicago: University Of Chicago Press, 1987.

Johnson, Sarah Kathleen. "Online Communion, Christian Community, and Receptive Ecumenism: A Holy Week Ethnography during COVID-19." *Studia Liturgica* 50, no. 2 (2020): 188–210.

Jonas, Hans. *The Imperative of Responsibility: In Search of an Ethics for the Technological Age*. Translated by Hans Jonas and David Herr. Chicago: University of Chicago Press, 1984.

Joustra, Jessica. "An Embodied *Imago Dei*: How Herman Bavinck's Understanding of the Image of God Can Help Inform Conversations on Race." *Journal of Reformed Theology* 11 (2017): 9–23.

Jung, Alexander. *Machine Learning: The Basics*. Singapore: Springer, 2022.

Junius, Franciscus. "The Life of the Noble and Learned Franciscus Junius, Doctor of Sacred Theology and Distinguished Professor at Leiden University." In *A Treatise on True Theology: With the Life of Franciscus Junius*. 1–72. Grand Rapids: Reformation Heritage Books, 2014.

Junius, Franciscus. *A Treatise on True Theology: With the Life of Franciscus Junius*. Translated by David C. Noe. Grand Rapids: Reformation Heritage Books, 2014.

Kaunda, Chammah Judex. "Bemba Mystico-Relationality and the Possibility of Artificial General Intelligence (AGI) Participation in *Imago Dei*." *Zygon: Journal of Religion and Science* 55, no. 2 (2020): 327–343.

Kelsey, David H. *Eccentric Existence: A Theological Anthropology*. 2 vols. Louisville, Ky.: Westminster John Knox, 2009.

Kerr, Ian R. "Spirits in the Material World: Intelligent Agents as Intermediaries in Electronic Commerce." *Dalhousie Law Journal* 22, no. 2 (1999): 190–249.

Kilmartin, Edward J. *Systematic Theology of Liturgy*. Christian Liturgy: Theology and Practice, Volume I. Kansas: Sheed & Ward, 1988.

Knifka, Julia. "On the Significance of Understanding in Human–Robot Interaction." In *Social Robots: Boundaries, Potential, Challenges*. Edited by Marco Nørskov, 3–17. Farnham: Ashgate, 2016.

Knight, Christopher C. *The God of Nature: Incarnation and Contemporary Science*. Minneapolis: Fortress, 2007.

Kuipers, Benjamin. "Drinking from the Firehose of Experience." *Artificial Intelligence in Medicine* 44 (2008): 155–170.

Kurzweil, Raymond. *The Singularity Is Near: When Humans Transcend Biology*. London: Viking Penguin, 2005.

Kurzweil, Raymond. *The Age of Intelligence Machines*. Cambridge: MIT, 1990.

Kurzweil, Raymond. *The Age of Spiritual Machines: When Computers Exceed Human Intelligence*. New York: Viking, 1999.

Kuyper, Abraham. *Our Worship*. Translated by Harry Boonstra, Henry Baron, Gerrit Sheeres and Leonard Sweetman. Edited by John D. Witvliet and Harry Boonstra. Grand Rapids: Eerdmans, 2009.

Lamm, Claus, Jean Decety, and Tania Singer. "Meta-Analytic Evidence for Common and Distinct Neural Networks Associated with Directly Experienced Pain and Empathy for Pain." *NeuroImage* 54 (2011): 2492–2502.

Landgrebe, Jobst, and Barry Smith. *Why Machines Will Never Rule the World: Artificial Intelligence Without Fear*. New York: Routledge, 2023.

Lane, Justin E. "Method, Theory, and Multi-Agent Artificial Intelligence: Creating Computer Models of Complex Social Interaction." *Journal for the Cognitive Science of Religion* 1, no. 2 (2013): 161–180.

Lane, Justin E. *Understanding Religion through Artificial Intelligence: Bonding and Belief*. London: Bloomsbury Academic, 2021.

Levy, David. *Robots Unlimited: Life in a Virtual Age*. Wellesley, Massachusetts: A K Peters, 2006.

Levy, David. *Love and Sex with Robots: The Evolution of Human-Robot Relationships*. New York: HarperCollins, 2008.

Levy, David. "Falling in Love with a Companion." In *Close Engagements with Artificial Companions: Key Social, Psychological, Ethical and Design Issues*. Edited by Yorick Wilks, 89–93. Amsterdam: John Benjamins, 2010.

Lim, Chee-Peng, Ashlesha Vaidya, Kiran Jain, Virag U. Mahorkar, and Lakhmi Jain, eds. *Handbook of Artificial Intelligence in Healthcare, Vol. 1—Advances and Applications*. Cham: Springer, 2022.

Littman, Michael L., Ifeoma Ajunwa, Guy Berger, Craig Boutilier, Morgan Currie, Finale Doshi-Velez, Gillian Hadfield, *et al. Gathering Strength, Gathering Storms: The One Hundred Year Study on Artificial Intelligence (AI100) 2021 Study Panel Report*. Stanford University (Stanford, CA: September 2021). http://ai100.stanford.edu/2021-report.

Liu, Yuxin, Adam Moore, Jamie Webb, and Shannon Vallor. "Artificial Moral Advisors: A New Perspective from Moral Psychology." Paper presented at the Proceedings of 2022 ACM/AAAI Conference on AI, Ethics, and Society (AIES '22), Oxford, August 1–3, 2022.

Longinotti, David. "Agency, Qualia and Life: Connecting Mind and Body Biologically." In *Philosophy and Theory of Artificial Intelligence 2017*. Edited by Vincent C. Müller, 43–56. Cham: Springer, 2018.

Lumbreras, Sara. "Lessons from the Quest for Artificial Consciousness: The Emergence Criterion, Insight-Oriented AI, and Imago Dei." *Zygon: Journal of Religion and Science* 57, no. 4 (2022): 963–983.

Luxton, David D., and Eva Hudlicka. "Intelligent Virtual Agents in Behavioral and Mental Healthcare: Ethics and Application Considerations." In *Artificial Intelligence in Brain and Mental Health: Philosophical, Ethical & Policy Issues*. Edited by Fabrice Jotterand and Marcello Lenca, 53–55. Cham: Springer, 2021.

Lyall, David. *The Integrity of Pastoral Care*. London: SPCK, 2001.

Lydon, Cora. "Edinburgh Hospital Trials AI-powered Pain Assessment Tech." DigitalHealth, 21 December 2023, accessed 30 December, 2023, https://www.digitalhealth.net/2023/12/edinburgh-hospital-trials-ai-powered-pain-assessment-tech/.

Macinnis, Adam. "An AI Aims to Be First Christian Celebrity of the Metaverse." *Christianity Today*, 14 February 2022. https://www.christianitytoday.com/ct/2022/march/ai-artificial-intelligence-music-metaverse-gospel-singer-jc.html.

Manson, Neil. "State Consciousness and Creature Consciousness: A Real Distinction." *Philosophical Psychology* 13, no. 3 (2000): 405–410.

Manzotti, Riccardo, and Sabina Jeschke. "A Causal Foundation for Consciousness in Biological and Artificial Agents." *Cognitive Systems Research* 40 (2016): 172–185.

Manzotti, Riccardo, and Vincenzo Tagliasco. "Artificial Consciousness: A Discipline between Technological and Theoretical Obstacles." *Artificial Intelligence in Medicine* 44 (2008): 105–117.

Marchant, Gary, and Lucille Nalbach Tournas. "Matter Over Mind: Liability Considerations Surrounding Artificial Intelligence in Neuroscience." In *Artificial Intelligence in Brain and Mental Health: Philosophical, Ethical & Policy Issues*. Edited by Fabrice Jotterand and Marcello Lenca, 233–246. Cham: Springer, 2021.

Matthias, Andreas. "The Responsibility Gap: Ascribing Responsibility for the Actions of Learning Automata." *Ethics and Information Technology* 6 (2004): 175–183.

Mattson, Brian G. *Restored to Our Destiny: Eschatology & the Image of God in Herman Bavinck's Reformed Dogmatics*. Leiden: Brill, 2012.

McCorduck, Pamela. *Machines Who Think: A Personal Inquiry into the History and Prospects of Artificial Intelligence*. Natick, Massachusetts: A K Peters, 2004.

McDermott, Drew. "Artificial Intelligence and Consciousness." In *The Cambridge Handbook of Consciousness*. Edited by Philip David Zelazo, Morris Moscovitch and Evan Thompson, 117–150. Cambridge: Cambridge University Press, 2007.

McDonald, Erin M., Katrina D. Farris, Arden M. Cooper, Meghan Rose Donohue, and Erin C. Tully. "Activity Increase in Empathy-Related Brain Regions When Children Contribute to Peers' Sadness and Happiness." *Journal of Experimental Child Psychology* 239 (2024): 105812.

McFadyen, Alistair. "Redeeming the Image." *International Journal for the Study of the Christian Church* 16, no. 2 (2016): 108–125.

McFarland, Ian A. *The Word Made Flesh: A Theology of the Incarnation*. Louisville: Westminster John Knox, 2019.

Mercer, Calvin, and Tracy J. Trothen. *Religion and the Technological Future: An Introduction to Biohacking, Artificial Intelligence, and Transhumanism*. Cham: Palgrave Macmillan, 2021.

Midson, Scott A. "From *Imago Dei* to Social Media: Computers, Companions and Communities." In *Love, Technology and Theology*. Edited by Scott A. Midson, 143–160. London: T&T Clark, 2020.

Mohri, Mehryar, Afshin Rostamizadeh, and Ameet Talwalkar. *Foundations of Machine Learning*. 2nd ed. Cambridge, Massachusetts: MIT Press, 2018.

Moltmann, Jürgen. *Ethics of Hope*. Translated by Margaret Kohl. Minneapolis: Fortress Press, 2012.

Monsma, Stephen V., Clifford Christians, Eugene R. Dykema, Arie Leegwater, Egbert Schuurman, and Lambert Van Poolen. *Responsible Technology: A Christian Perspective*. Edited by Stephen V. Monsma. Grand Rapid: Eerdmans, 1986.

Moor, James H. "Four Kinds of Ethical Robots." *Philosophy Now* 72 (2009): 12–14.

Moor, James H. "The Nature, Importance, and Difficulty of Machine Ethics." *IEEE Intelligent Systems* 21, no. 4 (2006): 18–21.

Moravec, Hans P. *Mind Children: The Future of Robot and Human Intelligence*. Cambridge, Massachusetts: Harvard University Press, 1988.

Morrill, Bruce T. *Anamnesis as Dangerous Memory: Political and Liturgical Theology in Dialogue*. Collegeville: Liturgical Press, 2000.

Muller, Richard A. *Post-Reformation Reformed Dogmatics, Volume One: Prolegomena to Theology*. 2nd ed. Grand Rapids: Baker, 2003.

Murphy, Nancey. *Bodies and Souls, or Spirited Bodies?* Cambridge: Cambridge University Press, 2006.

Murphy, Nancey. "Nonreductive Physicalism: Philosophical Issues." In *Whatever Happened to the Soul?: Scientific and Theological Portraits of Human Nature*. Edited by Warren S. Brown, Nancey Murphy, and H. Newton Malony, 127–148. Minneapolis: Fortress, 1998.

Murphy, Nancey. "Physicalism Without Reductionism: Toward a Scientifically, Philosophically, and Theologically Sound Portrait of Human Nature." *Zygon: Journal of Religion and Science* 34, no. 4 (1999): 551–571.

Newell, Allen. "The Knowledge Level: Presidential Address." *AI Magazine* 2, no. 2 (1981): 1–33.

Niebuhr, Reinhold. *The Nature and Destiny of Man*. 2 vols. Louisville: Westminster John Knox, 1996.

Noble, David F. *The Religion of Technology: The Divinity of Man and the Spirit of Invention*. New York: Alfred A. Knopf, 1998.

O'Donnell, Karen. "Performing the *Imago Dei*: Human Enhancement, Artificial Intelligence and Optative Image-Bearing." *International Journal for the Study of the Christian Church* 18, no. 1 (2018): 4–15.

O'Donovan, Oliver. *Begotten or Made?* Oxford: Clarendon, 1984.

Parry, Robin A. *Worshipping Trinity: Coming Back to the Heart of Worship*. 2nd ed. Eugene: Cascade Books, 2012.

Peterson, Gregory R. "The Created Co-Creator: What It Is and Is Not." *Zygon: Journal of Religion and Science* 39, no. 4 (2004): 827–840.

Peterson, Gregory R. "Is My Feeling Your Pain Bad for Others? Empathy as Virtue versus Empathy as Fixed Trait." *Zygon: Journal of Religion & Science* 52, no. 1 (2017): 232–257.

Pfeifer, Rolf, and Fumiya Iida. "Embodied Artificial Intelligence: Trends and Challenges." In *Embodied Artificial Intelligence*. Edited by Fumiya Iida, Rolf Pfeifer, Luc Steels and Yasuo Kuniyoshi, 1–26. Berlin: Springer, 2004.

Piccinini, Gualtiero. "The Ontology of Creature Consciousness: A Challenge for Philosophy." *Behavioral and Brain Sciences* 30, no. 1 (2007): 103–104.

Pineau, Joelle, Michael Montemerlo, Martha Pollack, Nicholas Roy, and Sebastian Thrun. "Towards Robotic Assistants in Nursing Homes: Challenges and Results." *Robotics and Autonomous Systems* 42 (2003): 271–281.

Postman, Neil. *Technopoly: The Surrender of Culture to Technology*. New York: Vintage, 1993.

Powers, Thomas M., and Jean-Gabriel Ganascia. "The Ethics of the Ethics of AI." In *The Oxford Handbook of Ethics of AI*. Edited by Markus D. Dubber, Frank Pasquale and Sunit Das, 27–51. Oxford: Oxford University Press, 2020.

Preston, John, and Mark Bishop, eds. *Views into the Chinese Room: New Essays on Searle and Artificial Intelligence*. Oxford: Clarendon, 2002.

Preston, Stephanie D., and Frans B. M. de Waal. "Empathy: Its Ultimate and Proximate Bases." *Behavioral and Brain Sciences* 25 (2002): 1–72.

Preus, Robert D. *A Study of Theological Prolegomena*. The Theology of Post-Reformation Lutheranism, Volume I. St. Louis, Missouri: Concordia Publishing House, 1970.

Puddefoot, John C. *God and the Mind Machine: Computers, Artificial Intelligence, and the Human Soul*. London: SPCK, 1996.

Quitterer, Josef. "Liturgical Rituals as Shared Intentional Practices." *Studia Liturgica* 48, no. 1–2 (2018): 56–67.

Ratzinger, Joseph. *The Spirit of the Liturgy*. Translated by John Saward. San Francisco: Ignatius Press, 2000.

"Responsibility Projects," UKRI Trustworthy Autonomous Systems Hub, accessed by 6 February 2024, https://tas.ac.uk/responsibility-projects/.

Ritchie, Sarah Lane. *Divine Action and the Human Mind*. Cambridge: Cambridge University Press, 2019.

Rizzolatti, Giacomo. "The Mirror Neuron System and Its Function in Humans." *Anatomy and Embryology* 210, no. 5–6 (2005): 419–421.

Rizzolatti, Giacomo, and Laila Craighero. "The Mirror-Neuron System." *Annual Review of Neuroscience* 27 (2004): 169–192.

Russell, Stuart. *Human Compatible: Artificial Intelligence and the Problem of Control*. New York: Viking, 2019.

Schleiermacher, Friedrich. *On Religion: Speeches to Its Cultured Despisers*. Edited and translated by Richard Crouter. 2nd ed. Cambridge: Cambridge University Press, 1996.

Schmidt, Katherine G. *Virtual Communion: Theology of the Internet and the Catholic Sacramental Imagination*. Lanham: Lexington Books, 2020.

Schuurman, Derek C. "Artificial Intelligence: Discerning a Christian Response." *Perspectives on Science and Christian Faith* 71, no. 2 (2019): 75–82.

Seager, William. "A Brief History of the Philosophical Problem of Consciousness." In *The Cambridge Handbook of Consciousness*. Edited by Philip David Zelazo, Morris Moscovitch and Evan Thompson, 9–33. Cambridge: Cambridge University Press, 2007.

Searle, John R. "How to Study Consciousness Scientifically." In *Consciousness and Human Identity*. Edited by John Cornwell. Oxford: Oxford University Press, 1998.

Searle, John R. *Intentionality: An Essay in the Philosophy of Mind*. Cambridge: Cambridge University Press, 1983.

Searle, John R. *Mind: A Brief Introduction*. Oxford: Oxford University Press, 2004.

Searle, John R. *Minds, Brains and Science*. Cambridge, Massachusetts: Harvard University Press, 1984.

Searle, John R. "Minds, Brains, and Programs." *The Behavioral and Brain Sciences* 3 (1980): 417–457.

Sekmen, Ali, and Prathima Challa. "Assessment of Adaptive Human-Robot Interactions." *Knowledge-Based Systems* 42 (2013): 49–59.

Senn, Frank C. *Christian Liturgy: Catholic and Evangelical*. Minneapolis: Fortress, 1997.

Senn, Frank C. *Introduction to Christian Liturgy*. Minneapolis: Fortress, 2012.

Senn, Frank C. *Embodied Liturgy: Lessons in Christian Ritual*. Minneapolis: Fortress, 2016.

Sharkey, Amanda. "Can Robots Be Responsible Moral Agents? And Why Should We Care?." *Connection Science* 29, no. 3 (2017): 210–216.

Sharkey, Amanda, and Noel Sharkey. "Granny and the Robots: Ethical Issues in Robot Care for the Elderly." *Ethics and Information Technology* 14, no. 1 (2012): 27–40.

Sharkey, Noel, and Amanda Sharkey. "Living with Robots: Ethical Tradeoffs in Eldercare." In *Close Engagements with Artificial Companions: Key Social, Psychological, Ethical and Design Issues*. Edited by Yorick Wilks, 245–255. Amsterdam: John Benjamins, 2010.

Sigurdson, Ola. *Heavenly Bodies: Incarnation, the Gaze, and Embodiment in Christian Theology*. Translated by Carl Olsen. Grand Rapids: Eerdmans, 2016.

Singler, Beth. "An Introduction to Artificial Intelligence and Religion for the Religious Studies Scholar." *Implicit Religion* 20, no. 3 (2017): 215–231.

Sison, Alejo José G., and Dulce M. Redin. "A Neo-Aristotelian Perspective on the Need for Artificial Moral Agents (AMAs)." *AI & Society* 38 (2023): 47–65.

Sloman, Aaron. "Requirements for Artificial Companions: It's Harder than You Think." In *Close Engagements with Artificial Companions: Key Social, Psychological, Ethical and Design Issues*. Edited by Yorick Wilks, 179–200. Amsterdam: John Benjamins, 2010.

Sloman, Aaron, and Ron Chrisley. "Virtual Machines and Consciousness." *Journal of Consciousness Studies* 10, no. 4–5 (2003): 133–172.

Slotnick, Scott D. *Cognitive Neuroscience of Memory*. Cambridge: Cambridge University Press, 2017.

Slotnick, Scott D., and Daniel L. Schachter. "The Cognitive Neuroscience of Memory and Consciousness." In *The Cambridge Handbook of Consciousness*. Edited by Philip David Zelazo, Morris Moscovitch and Evan Thompson, 809–827. Cambridge: Cambridge University Press, 2007.

Song, Yong Sup. "Religious AI as an Option to the Risks of Superintelligence: A Protestant Theological Perspective." *Theology and Science* 19, no. 1 (2021): 65–78.

Sparrow, Robert. "Why Machines Cannot Be Moral." *AI & Society* 36 (2021): 685–693.

Sparrow, Robert, and Linda Sparrow. "In the Hands of Machines? The Future of Aged Care." *Minds and Machines* 16 (2006): 141–161.

Spinello, Richard A. "Karol Wojtyla on Artificial Moral Agency and Moral Accountability." *The National Catholic Bioethics Quarterly* 11, no. 3 (2011): 469–501.

Srinivasan, Ramya, and Beatriz San Miguel González. "The Role of Empathy for Artificial Intelligence Accountability." *Journal of Responsible Technology* 9 (2022): 1–7.

Stevens, Francis, and Katherine Taber. "The Neuroscience of Empathy and Compassion in Pro-social Behavior." *Neuropsychologia* 159 (2021): 107925.

Stewart, Jacqui. "Technology and Christianity." In *God, Humanity and the Cosmos: A Textbook in Science and Religion*. Edited by Christopher Southgate, 371–389. London: T&T Clark, 2011.

Stoddart, Eric. "Artificial Pastoral Care: Abdication, Delegation or Collaboration?." *Studies in Christian Ethics* 36, no. 3 (2023): 660–674.

Stone, Peter, Rodney Brooks, Erik Brynjolfsson, Ryan Calo, Oren Etzioni, Greg Hager, Julia Hirschberg, *et al. Artificial Intelligence and Life in 2030: One Hundred Year Study on Artificial Intelligence 2015 Study Panel Report*. Stanford University (Stanford: September 2016). https://ai100.stanford.edu/sites/g/files/sbiybj18871/files/media/file/ai100report10032016fnl_singles.pdf.

Strathearn, Carl, and Minhua Ma. "Modelling User Preference for Embodied Artificial Intelligence and Appearance in Realistic Humanoid Robots." *Informatics* 7, no. 3 (2020): 28.

Sullins, John. "Artificial Moral Agency in Technoethics." In *Handbook of Research on Technoethics*. Edited by Rocci Luppicini and Rebecca Adell, 205–221. Hershey: Information Science Reference, 2009.

Sullins, John. "When Is a Robot a Moral Agent?" In *Machine Ethics*. Edited by Michael Anderson and Susan Leigh Anderson, 151–161. Cambridge: Cambridge University Press, 2011.

Sun, Ron. "Connectionism and Neural Networks." In *The Cambridge Handbook of Artificial Intelligence*. Edited by Keith Frankish and William M. Ramsey, 108–127. Cambridge: Cambridge University Press, 2014.

Sziron, Mónika. "A Common Ground for Human Rights, AI, and Brain and Mental Health." In *Artificial Intelligence in Brain and Mental Health: Philosophical, Ethical & Policy Issues*. Edited by Fabrice Jotterand and Marcello Lenca, 247–258. Cham: Springer, 2021.

Te Velde, Dolf, ed. *Synopsis of a Purer Theology, Volume 1*. Leiden: Brill, 2015.

Teilhard de Chardin, Pierre. *The Future of Man*. Translated by Norman Denny. New York: Image Books, 2004.

Terada, Kazunori, Takashi Shamoto, and Akira Ito. "Human Goal Attribution toward Behavior of Artifacts." In *17th IEEE International Symposium on Robot and Human Interactive Communication*. 160–165: IEEE, 2008.

Terrighena, Esslin L., and Tatia M.C. Lee. "The Neuroimaging of Vicarious Pain." In *Neuroimaging of Pain*. Edited by Luca Saba, 411–451. Cham: Springer, 2017.

Tertullian. "The Prescription against Heretics." In *Latin Christianity: Its Founder, Tertullian*. Edited by Alexander Roberts, James Donaldson and A. Cleveland Coxe. Translated by Peter Holmes. Ante-Nicene Fathers, vol. III. Buffalo: Christian Literature Company, 1985.

Teske, John A. "From Embodied to Extended Cognition." *Zygon: Journal of Religion & Science* 48, no. 3 (2013): 759–787.

The Doctrine Commission of the Church of England. *We Believe in God*. London: Church House, 1987.

The Church of Scotland. *Artificial Intelligence (AI): Opportunity and Challenge for the Church*. (May 2021).

"The UKRI Trustworthy Autonomous Systems (TAS) Hub," UKRI Trustworthy Autonomous Systems Hub, accessed by 6 February 2024, https://tas.ac.uk/.

Thompson, Richard F., and Stephen A. Madigan. *Memory: The Key to Consciousness*. Princeton: Princeton University Press, 2005.

Thomson, Ann. "Materialistic Theories of Mind and Brain." In *Between Leibniz, Newton, and Kant: Materialistic Theories of Mind and Brain*. Edited by Wolfgang Lefèvre, 149–173. Dordrecht: Springer, 2001.

Torevell, David. *Losing the Sacred: Ritual, Modernity, and Liturgical Reform*. London: T&T Clark, 2004.

Torrance, Steve. "Ethics, Mind and Artifice." In *Artificial Intelligence for Society*. Edited by Karamjit S. Gill, 55–72. Somerset: Wiley-Blackwell, 1986.

Torrance, Steve. "Ethics and Consciousness in Artificial Agents." *AI & Society* 22 (2008): 495–521.

Torrance, Steve. "Artificial Consciousness and Artificial Ethics: Between Realism and Social Relationism." *Philosophy and Technology* 27, no. 1 (2014): 9–29.

Torrance, Thomas F. *Space, Time and Incarnation*. Edinburgh: T&T Clark, 1997.

Tseng, Shao Kai. *Barth's Ontology of Sin and Grace: Variations on a Theme of Augustine*. London: Routledge, 2019.

Turing, Alan M. "Computing Machinery and Intelligence." *Mind: A Quarterly Review of Psychology and Philosophy* 59, no. 236 (1950): 433–460.

Turing, Alan M. "Computing Machinery and Intelligence." In *Parsing the Turing Test: Philosophical and Methodological Issues in the Quest for the Thinking Computer*. Edited by Robert Epstein, Gary Roberts, and Grace Beber, 23–66. Dordrecht: Springer, 2009.

Turkle, Sherry. *Alone Together: Why We Expect More from Technology and Less from Each Other*. New York: Basic Books, 2011.

Turkle, Sherry. *The Second Self: Computers and the Human Spirit*. 20th Anniversary ed. Cambridge, Massachusetts: MIT Press, 2005.

Turretin, Francis. *Institutes of Elenctic Theology*. Translated by George Musgrave Giger. Edited by James T. Dennison Jr. 3 vols. Phillipsburg: P&R Publishing, 1992–1997.

Tye, Michael. "Philosophical Problems of Consciousness." In *The Blackwell Companion to Consciousness*. Edited by Susan Schneider and Max Velmans, 17–31. Chichester: Wiley Blackwell, 2017.

Ulanowicz, Robert E. "The Universal Laws of Physics: Inflated Ontologies?" In *God's Providence and Randomness in Nature: Scientific and Theological Perspectives*. Edited by Robert John Russell and Joshua M. Moritz, 69–84. West Conshohocken: Templeton Press, 2018.

Vallor, Shannon. "Carebots and Caregivers: Sustaining the Ethical Ideal of Care in the Twenty-First Century." *Philosophy & Technology* 24, no. 3 (2011): 251–268.

Vallor, Shannon. "Edinburgh Declaration on Responsibility for Responsible AI." 14 July 2023, accessed 7 September, 2023, https://medium.com/@svallor_10030/edinburgh-declaration-on-responsibility-for-responsible-ai-1a98ed2e328b#:~:text='%20All%20who%20exercise%20and%20unleash,%2C%20perpetuate%2C%20amplify%20or%20relieve.

Vallor, Shannon. *Technology and the Virtues: A Philosophical Guide to a Future Worth Wanting*. Oxford: Oxford University Press, 2016.

Van Asselt, Willem J. "The Fundamental Meaning of Theology: Archetypal and Ectypal Theology in Seventeenth-Century Reformed Thought." *Westminster Theological Journal* 64, no. 2 (2002): 319–335.

Van Asselt, Willem J. "Introduction." In *A Treatise on True Theology: With the Life of Franciscus Junius*. xi–xlvii. Grand Rapids: Reformation Heritage Books, 2014.

Van Asselt, Willem J. "Scholasticism in the Time of Early Orthodoxy (Ca. 1560–1620)." In *Introduction to Reformed Scholasticism*. Translated by Gootjes, Albert, 103–131. Grand Rapids: Reformation Heritage Books, 2011.

Van den Brink, Gijsbert. *Reformed Theology and Evolutionary Theory*. Grand Rapids: Eerdmans, 2020.

Van Huyssteen, J. Wentzel. *Alone in the World? Human Uniqueness in Science and Theology*. Grand Rapids: Eerdmans, 2006.

Van Huyssteen, J. Wentzel. "Questions, Challenges, and Concerns for the Image of God." In *Finding Ourselves after Darwin: Conversations on the Image of God, Original Sin, and the Problem of Evil*. Edited by Stanley P. Rosenberg, Michael Burdett, Michael Lloyd, and Benno Van den Toren, 33–47. Grand Rapids: Baker Academic, 2018.

Van Urk-Coster, Eva. "Created in the Image of God: Both Human and Non-Human Animals?" *Theology and Science* 19, no. 4 (2021): 343–362.

Van Wynsberghe, Aimee. "Designing Robots for Care: Care Centred Value-Sensitive Design." *Science and Engineering Ethics* 19, no. 2 (2013): 407–433.

Van Wynsberghe, Aimee. *Healthcare Robots: Ethics, Design and Implementation*. London: Routledge, 2016.

Van Wynsberghe, Aimee. "Care Ethics, Philosophy of Technology, and Robots in Humanitarian Action." In *The Oxford Handbook of Philosophy of Technology*. Edited by Shannon Vallor, 629–646. Oxford: Oxford University Press, 2021.

Van Wynsberghe, Aimee, and Scott Robbins. "Critiquing the Reasons for Making Artificial Moral Agents." *Science and Engineering Ethics* 25, no. 3 (2019): 719–735.

Von Allmen, Jean-Jacques. *Worship: Its Theology and Practice*. New York: Oxford University Press, 1965.

Wainwright, Geoffrey. "Christian Worship: Scriptural Basis and Theological Frame." In *The Oxford History of Christian Worship*. Edited by Geoffrey Wainwright and Karen B. Westerfield Tucker, 1–31. Oxford: Oxford University Press, 2005.

Wallach, Wendell, and Shannon Vallor. "Moral Machines: From Value Alignment to Embodied Virtue." In *Ethics of Artificial Intelligence*. Edited by S. Matthew Liao, 383–412. Oxford: Oxford University Press, 2020.

Wang, Pei, and Ben Goertzel. "Introduction: What Is the Matter Here?" In *Theoretical Foundations of Artificial General Intelligence*. Edited by Pei Wang and Ben Goertzel, 1–9. Paris: Atlantis, 2012.

Watkins, Eric, ed. *Kant on Persons and Agency*. Cambridge: Cambridge University Press, 2018.

Watson, Gary. "Moral Agency." In *The International Encyclopedia of Ethics*. Edited by Hugh LaFollette. 3322–3333. Oxford: Wiley-Blackwell, 2013. https://doi-org.ezproxy.is.ed.ac.uk/10.1002/9781444367072.wbiee294.

Watts, Fraser. "Artificial Intelligence." In *God, Science and Humility: Ten Scientists Consider Humility Theology*. Edited by Robert L. Herrmann, 279–300. Philadelphia: Templeton Foundation Press, 2000.

Watts, Fraser. "Embodied Cognition and Religion." *Zygon: Journal of Religion & Science* 48, no. 3 (2013): 745–758.

Watts, Fraser. *A Plea for Embodied Spirituality: The Role of the Body in Religion*. London: SCM, 2021.

Watts, Fraser, and Mark Williams. "The Doctrine Commission and Artificial Intelligence." *Theology* 91, no. 741 (1988): 216–217.

Weber, Jutta. "Opacity Versus Computational Reflection: Modelling Human-Robot Interaction in Personal Service Robotics." *Science, Technology & Innovation Studies* 10, no. 1 (2014): 187–199.

Weizenbaum, Joseph. "ELIZA—A ComputerProgram for the Study of Natural Language Communication Between Man And Machine." *Communications of the ACM* 9, no. 1 (1966): 36–45.

Wentzer, Thomas Schwarz. "Approaching Philosophical Anthropology: Human, the Responsive Being." In *Finite but Unbounded: New Approaches in Philosophical Anthropology*. Edited by Kevin M. Cahill, Martin Gustafsson, and Thomas Schwarz Wentzer, 25–45. Berlin: De Gruyter, 2017.

Westermann, Claus. *Genesis 1-11: A Continental Commentary*. Minneapolis: Fortress, 1994.

Whitby, Blay. "Computing Machinery and Morality." *AI & Society* 22 (2008): 551–563.

White, Susan. *Christian Worship and Technological Change*. Nashville: Abingdon, 1994.

Wiegel, Vincent. "The Ethics of IT-Artefacts." In *The Cambridge Handbook of Information and Computer Ethics*. Edited by Luciano Floridi, 201–218. Cambridge: Cambridge University Press, 2010.

Wilks, Yorick. "Foreword." In *Close Engagements with Artificial Companions: Key Social, Psychological, Ethical and Design Issues*. Edited by Yorick Wilks, xi–xii. Amsterdam: John Benjamins, 2010.

Williams, Rowan. "Making Moral Decisions." In *The Cambridge Companion to Christian Ethics*. Edited by Robin Gill, 3–15. Cambridge: Cambridge University Press, 2001.

Wiseman, Harris. "Meaning and Embodiment in Ritual Practice." *Zygon: Journal of Religion and Science* 57, no. 3 (2022): 772–796.

Wiseman, Harris. "The Sins of Moral Enhancement Discourse." *Royal Institute of Philosophy Supplement* 83 (2018): 35–58.

Wolterstorff, Nicholas. *Acting Liturgically: Philosophical Reflection on Religious Practice*. Oxford: Oxford University Press, 2018.

Wolterstorff, Nicholas. *The God We Worship: An Exploration of Liturgical Theology*. Grand Rapids: Eerdmans, 2015.

Wolterstorff, Nicholas. "Knowing God Liturgically." *Journal of Analytic Theology* 4 (2016): 1–16.

Woodruff, C. Chad. "Reflections of Others and of Self: The Mirror Neuron System's Relationship to Empathy." In *The Neuroscience of Empathy, Compassion, and Self-Compassion*. Edited by Larry Stevens and C. Chad Woodruff, 157–187. London: Academic Press, 2018.

Wright, James, David Leslie, Charles Raab, Fumi Kitagawa, Florian Ostmann, and Morgan Briggs. *Privacy, Agency and Trust in Human-AI Ecosystems: Interim Report (Short Version)*. London: The Alan Turing Institute, 2021.

Xu, Ximian. "A Theological Account of Artificial Moral Agency." *Studies in Christian Ethics* 36, no. 3 (2023): 642–659.

Xu, Ximian. "Karl Barth's Ontology of Holy Scripture Revisited." *Scottish Journal of Theology* 74, no. 1 (2021): 26–40.

Young, William. "Reverend Robot: Automation and Clergy." *Zygon: Journal of Religion and Science* 54, no. 2 (2019): 479–500.

Young, William. "Virtual Pastor: Virtualization, AI, and Pastoral Care." *Theology and Science* 20, no. 1 (2022): 6–22.

Index

Made in the USA
Las Vegas, NV
19 September 2026

0b847faa-3880-40ae-ab1f-7e8eab3b74a9R01